The Honey Bee and Climate Change

The Honey Bee and Climate Change

Gloria DeGrandi-Hoffman
Retired, USDA, Agricultural Research Service, College of Integrative Sciences and Arts, School of Applied Sciences and Arts, Arizona State University, Tempe, Arizona, United States

ELSEVIER

ACADEMIC PRESS
An imprint of Elsevier

Academic Press is an imprint of Elsevier
125 London Wall, London EC2Y 5AS, United Kingdom
50 Hampshire Street, 5th Floor, Cambridge, MA 02139, United States

For accessibility purposes, images in electronic versions of this book are accompanied by alt text descriptions independently provided by Elsevier, generated with the aid of AI. For more information, see https://www.elsevier.com/about/accessibility or Radarweg 29, 1043 NX Amsterdam, Netherlands.

Books and Journals published by Elsevier comply with applicable product safety requirements. For any product safety concerns or queries, please contact our authorised representative, Elsevier B.V., at productsafety@elsevier.com.

Publisher's note: Elsevier takes a neutral position with respect to territorial disputes or jurisdictional claims in its published content, including in maps and institutional affiliations.

Notices

Knowledge and best practice in this field are constantly changing. As new research and experience broaden our understanding, changes in research methods, professional practices, or medical treatment may become necessary.

Practitioners and researchers must always rely on their own experience and knowledge in evaluating and using any information, methods, compounds, or experiments described herein. In using such information or methods they should be mindful of their own safety and the safety of others, including parties for whom they have a professional responsibility.

To the fullest extent of the law, neither the Publisher nor the authors, contributors, or editors, assume any liability for any injury and/or damage to persons or property as a matter of products liability, negligence or otherwise, or from any use or operation of any methods, products, instructions, or ideas contained in the material herein.

ISBN: 978-0-443-28995-8

Publisher: Mara Conner
Acquisitions Editor: Kelsey Connors
Editorial Project Manager: Sweta Bharti
Production Project Manager: Sathyavani Deiva Sigamani
Cover Designer: Mark Rogers

Typeset by MPS Limited, Chennai, India

Dedication

To Rich, Mimi, Melissa, and Puppy. My favorite hiking buddies and the joy in my life.

Contents

About the author

Dr. Gloria DeGrandi-Hoffman is a retired supervisory research entomologist, research leader, and location coordinator at the Carl Hayden Bee Research Center, United States Department of Agriculture (USDA)–Agricultural Research Service (ARS) in Tucson, AZ, United States. She is an adjunct scientist in the Department of Entomology at the University of Arizona, Tucson, AZ, United States, and an adjunct faculty member in the School of Applied Sciences and Arts at Arizona State University, Tempe, AZ, United States. She has conducted research with honey bees and constructed mathematical models for over 40 years. She has studied crop pollination, Africanized honey bees, Varroa mites, impacts of fungicides, honey bee nutrition, and effects of climate change on colony health and survival. She has won numerous awards for her research, among them the John Henry Comstock Award from the Entomological Society of America, the J.I. Hambleton Award from the Eastern Apicultural Society, the Scientist of the Year Award from the USDA—Agricultural Research Service, the USDA Technology Transfer Award, and the Federal Laboratory Consortium Award, and she is an Alumni Fellow of the Pennsylvania State University. She continues to conduct research and build models to address the effects of climate change on honey bees and food security.

Preface

My interest in climate change and the effects on honey bee colonies began with two simultaneous yet unrelated events. Let me explain. The first was a conversation I had with my sister on Christmas Day. She lives on a farm in Pennsylvania and has some colonies of bees. She said it was a beautiful day, and she was sitting on the porch enjoying the unseasonably warm weather. In fact, it had been warm all week. I asked how her bees were doing, and she said they were right up there on the porch with her. I've spent years building population models to understand the dynamics of honey bee colonies and how they are affected by weather. One thing I knew for certain: bees in Pennsylvania need to stay in their hives during the winter and conserve their food and energy. They shouldn't be spending Christmas on the porch with my sister.

The weather over the Christmas holidays made me curious to see what warm temperatures in winter might do to colonies in temperate areas like Pennsylvania. As a rule, honey bees in temperate climates slow down and eventually stop rearing brood in the fall. During the cold of winter, the bees cluster together in the hive. The bees become active again in the spring as daytime temperatures become warmer.

Warmer falls and winters are predicted to occur more frequently due to climate change. Part of the reason is that summers are warmer, prolonging the time it takes to shift into fall patterns. To see what would happen to a colony of bees when there were warm winter temperatures, I ran my honey bee colony population model under weather conditions like those during that warm Christmastime in Pennsylvania. The model predicted that the warm temperatures would cause the bees to fly and the colony populations to decline sharply in the spring. Unless the colony was very large (more than 20,000 bees) when it went into winter, the model predicted that the population would probably dwindle and die in the spring even if the bees were otherwise healthy and had plenty of food.

The "bees on the porch" story has a second part. My realization that warm falls and winters could negatively affect colony survival in the spring occurred at the same time as I was part of a team developing a new miticide to control Varroa mites. Varroa is a parasite of honey bees and the leading cause of colony losses worldwide. During those studies, we sampled colonies and estimated Varroa populations throughout the year. From the estimates, it became apparent that numbers of mites were increasing far faster than we could explain by the mites reproducing in brood cells. We knew this because of my systems model that predicts honey bee colony and Varroa population growth due to reproduction. In the model, Varroa populations grow slowly because the mites have low reproductive rates. It takes about 2–3 years for populations that started out small to reach numbers where they threaten the colony's survival. The mite populations in our study were reaching abnormally high numbers in just a few months.

Whenever populations increase faster than expected by just reproducing, a possible explanation is that individuals are coming in from the outside, that is, immigration. There were previous reports of foraging bees carrying mites on their bodies when they came back to the hive from foraging trips. How frequently foragers with mites occurred and whether it affected Varroa numbers in a colony were not known.

We conducted studies to measure the frequency of finding foragers with mites at the entrances of colonies. The study began in the summer and continued through the fall. We found foraging bees carrying Varroa, especially during the fall, and it coincided with sharp increases in mite populations in the colonies. If bees are flying for longer periods in the fall because of warmer temperatures, this could extend the period when mites migrate into colonies on foragers. This means that colonies could go into winter with high mite populations even if Varroa populations were low all summer long. If warm falls and winters can cause colonies to decline in the spring because of extended foraging, and then mite migration and mortality from Varroa parasitism are added in, it is no wonder we were losing so many colonies over the winter!

The second unrelated event that deepened my understanding about the effects of climate change was listening to a conversation about the repercussions of Hurricane Katrina on communities in Louisiana (https://onbeing.org/programs/colette-pichon-battle-on-knowing-what-were-called-to/). A message from the podcast was that hurricanes did more than destroy property. They destroyed communities that had been on the same land for generations. Those communities contained connections among families and traditions that were tied to the land. Knowledge, stories, traditions, and culture passed down through generations were lost along with neighborhoods and property. Communities have history that includes babies being born and children growing, supporting each other through challenges and sharing joys and losses of loved ones. Many people never returned to the communities in and around New Orleans after Katrina, and an entire way of life dissolved. Losses of communities are being experienced across the country wherever climate-related disasters occur, and they are occurring with growing frequency.

So, what is the thread pulling together these simultaneous events about climate change? It was the realization that the effects of climate change are not just loss of property but a loss of connections that create communities.

The rendering is occurring on many levels, including species in ecosystems, bees and flowering plants, beekeeping operations, many of which are generational and family owned, and in human communities, contributing to a sense of isolation and disconnection.

Frequent large-scale losses of colonies have reduced the number of beekeepers in the United States. The thing about beekeeping is that it has had a long history in the United States, with beekeeping families going back for more than a century. There is a beekeeping community where everyone knows each other and has known their families for generations. Information and know-how are exchanged and passed on to the younger beekeepers. Colony losses go deeper than just the numbers, as they are also losses of beekeeping businesses and an erosion of beekeeping traditions and history.

Often the consequences of climate change are shown as extreme weather events like floods, hurricanes, and fires. However, climate change has subtle effects, and these are the primary focus of this book. Specifically, my focus is on the subtle effects of climate change on all pollinators, but particularly honey bees. I also discuss the broader implications of pollinator population decline on crop production and natural landscapes.

Honey bees are the primary pollinator of agricultural crops in many regions of the world. When beekeepers move hives into blooming fields and orchards, they provide large populations of pollinators that are needed to produce agricultural crops. About 75% of the human diet relies on honey bee pollination. A reliable source of honey bees is increasingly important to human welfare across the globe. Fruits and vegetables that are cornerstones of heart-healthy and cancer prevention diets are produced because of honey bee pollination. In fact, mortality rates from nonpathogenic diseases such as diabetes and cardiovascular illnesses can be prevented by increased consumption of foods that rely on honey bee pollination.

To understand the full effects of climate change requires a broad view of connections between and among organisms. Those connections are where the deepest disruptions are occurring. For this reason, I use a systems perspective to describe the effects of climate change and apply it to colony dynamics, pollination, and interactions between pathogens, especially Varroa mites, and colony survival. I also discuss climate change as the product of a system. With a systems perspective, there is an emphasis on the interconnectedness among individuals or on the components that make up the system. I frame the systems discussed in this book within the environmental conditions generated from climate change and discuss the repercussions.

I have a bias toward a systems perspective, as I have applied the techniques of systems science and mathematical modeling throughout my research career. I begin this book with an overview of systems science and what it means to think in systems. Systems are all around us, and we interact with them throughout the day. All organisms are comprised of incredibly complex systems. Historically, biologists have dealt with that complexity by studying single parts of organisms such as cells, molecules, and genes (i.e., reductionism). However, reducing systems into smaller parts and studying them intensely rarely reveals the workings of the system in its entirety. This applies to single organisms and to populations.

The theory of system science was introduced in 1937 by Ludwig von Bertalanffy as a counter approach to reductionism. Using a systems approach, organisms were studied as organized, interacting, dynamic subsystems rather than isolated parts. Likewise, ecosystems were seen as dynamic groups of interacting species. Systems thinking is needed to understand and address the challenges of climate change and colony losses. A systems approach requires looking at the whole picture, understanding all the parts, and recognizing that there might be multiple causes and solutions to problems. This approach moves us beyond symptoms to find root causes and the confluence of events that explain particular outcomes. Interdisciplinary collaborations are essential for understanding complex systems and for providing comprehensive solutions rather than quick fixes.

Throughout this book, I discuss honey bee colony dynamics from a systems perspective. The colony system is multidimensional in that the key to understanding honey bee colonies is to hold two thoughts in your mind simultaneously. The first is understanding the honey bee, especially worker bees, as individuals with their own physiological subsystems that enable them to perceive and respond to their environment. The second is understanding the colony as a whole system or organism with numerous interacting subsystems comprised of groups of worker bees participating in a common activity. There are numerous colony-level responses performed by groups of bees. Rearing a new queen, building wax comb, foraging, swarming, and nest defense are all group activities.

To appreciate how climate change can stress honey bees and cause colonies to perish, we first need to understand how a healthy colony operates throughout the year. I will explain how colonies go through an annual cycle that begins in late winter or early spring when bees start rearing new bees (i.e., brood). The colonies grow by rearing increasing amounts of brood throughout spring and summer. Brood rearing slows and, in some areas, stops in the fall as colonies prepare for winter. Colonies need about 100 lbs. (45.3 kg) of honey stored in the hive so that they will not starve if confined for long periods or if flowering plants are not available. To capture the complexities of the colony and how behaviors are coordinated so populations and food stores can grow, I describe different subsystems in colonies, including brood rearing, immunity, foraging, and thermoregulation. I chose these subsystems because they are the ones most clearly affected by climate change.

Just as I provide a description of subsystems in colonies, I describe plant subsystems that produce flowers, nectar, and pollen. I also describe how these subsystems have components that produce cues to attract pollinators and ensure that foraging for nectar and pollen leads to pollination and seed set.

Again, these are not the only subsystems in plants but are ones most directly affecting plant–pollinator relationships and that are being compromised by climate change.

To obtain a deeper appreciation of the extraordinary capabilities of plants, I begin the Chapter on pollination and fruit set with a discussion of photosynthesis. The system plants use to capture carbon dioxide (CO_2) and combine it with water and sunlight to produce simple sugars (glucose) is the engine for all life on earth. I also describe how plants are key components in systems to regulate levels of CO_2 in the atmosphere that ultimately influence ocean temperatures and currents that create weather patterns.

The systems that have kept CO_2 levels in check for more than 400 million years are being overwhelmed by the increasing use of fossil fuels. I describe the history of fossil fuels and how they drove industrial development from the mid-19th century to the present time. I include a historical perspective starting with the Industrial Revolution to convey how difficult the challenge will be to leave fossil fuels behind and move into noncarbon-based energy sources. Move on we must, though, as CO_2 levels from burning fossil fuels have repercussions on temperatures in the atmosphere, the oceans, and land through a phenomenon called "the greenhouse effect." CO_2 and other gases such as methane and water vapor are released into the atmosphere when fossil fuels are burned and absorb the sun's infrared radiation and trap heat like the glass panels of a greenhouse. Interestingly, we knew about the greenhouse effect from experiments conducted in the mid-19th century. Taking the long view, if there were an increasing use of fossil fuels and release of greenhouse gases, the impact on global atmospheric temperatures was an inevitable outcome. This takes us to where we are today, where burning fossil fuels creates a system of increasing atmospheric temperatures that leads to ocean warming, causing glaciers to melt and ocean levels to rise. Warmer oceans affect the frequency of hurricanes, typhoons, droughts, heatwaves, and changes in seasonal weather patterns that in turn influence flowering plants and their pollinators.

Massive losses of honey bee colonies first captured the public's attention during 2006–2007 when colony collapse disorder (CCD) became front-page news. The public quickly became aware of how important honey bees are to food production and how beekeepers were barely staying in business after losing extraordinary numbers of colonies. Though the causative agent for CCD was never identified, a contributing factor was determined to be Varroa mites. Varroa mites are external parasites of honey bees and can feed on immature and adult bees. To fully understand how Varroa successfully infests honey bee colonies, I begin with an overview of parasites in general. Parasites are among the most successful organisms on earth. Their success is largely due to their ability to adapt to current conditions, circumvent immune responses, use host behaviors to their advantage, and flee (i.e., disperse) and find new hosts when conditions are no longer favorable. Varroa follows general patterns for parasites and can disperse and infest colonies throughout apiaries and across landscapes. The mites disperse by attaching to foragers that then drift into other colonies. Varroa also harbors and transmits viruses. The viruses have made Varroa especially harmful to bees because even if miticides lower Varroa populations, the viruses in the bees continue to replicate and spread. Furthermore, when Varroa spreads to other colonies, viruses do too. I will explain the difficulties in controlling Varroa in colonies and how climate change has exacerbated the situation. Specifically, I will discuss how warmer fall temperatures extend the period when bees can forage and provide opportunities for Varroa to migrate and infest other colonies as they prepare for overwintering.

Throughout this book I discuss the importance of seeing a colony not just as a group of bees but as a population of individuals of different ages. Groups of bees that emerge as adults on the same day are an age group. The number of bees in each age group creates the colony age structure (i.e., numbers of worker bees that are 1-day old, 2-days old, 3-days old, etc.). The age structure depends on the amount of brood a colony rears and how long adult worker bees live. I will continue to refer to colony age structure throughout this book as it is the foundation for colony survival.

One of the subtle effects of climate change on honey bee colony health and survival is the effects on nutrition. As in humans and all other organisms, good nutrition is essential for maintaining health. Growth, development, reproduction, immunity, and numerous other vital functions depend on optimizing nutrition. Honey bees fulfill all their nutritional needs by consuming nectar and pollen. Nutritional needs of individual bees change throughout their lives. The nutritional needs of the colony also are dynamic and change throughout the year. The nutrients in pollen and nectar drive colony growth and survival because they determine how many eggs a queen will lay, how much brood a colony can rear, and how long adult worker bees will live. Climate change compromises honey bee nutrition by the effects that heat waves, drought, and flooding have on flowering plants and the nutrients in pollen and nectar.

Considering honey bee colonies from a systems perspective makes it apparent that the condition of a colony at any given time is the product of past events. When colonies are thriving in the spring, it is because they were large and healthy the previous fall. Likewise, if colonies are too small tosuccessfully overwinter, it is because events that occurred months before in the spring or summer curtailed colony growth when brood rearing should have been vigorous. Climate change can affect colony growth and survival during any time of the year. Warm temperatures early in the spring followed by freezes can stimulate brood rearing only to have flowering plants damaged and food sources eliminated literally overnight. Summer heat waves and droughts in some areas and flooding from tropical storms in others can reduce colony growth in the summer. Warm fall temperatures that prolong the period when bees can fly cause colonies to use food that should be stored for the winter. Flying also causes bees to physiologically age, so warm falls affect the age structure of the colony going into winter. These effects of warmer falls are in addition to providing more time for Varroa mites to spread among colonies.

Because adverse weather conditions are occurring with greater frequency and can happen at any time of year, it is becoming increasingly difficult for beekeepers to keep their colonies alive. Some colony losses are due to direct effects of climate change. These include flooding of apiaries from hurricanes and atmospheric rivers and fires from dry conditions. There also are indirect effects of climate change in addition to those on flowering plants and Varroa. These are the effects that higher temperatures have on pest insects. The rate that insects develop and grow their populations depends largely on temperature. The warmer the temperatures, the faster insects develop and populations grow. As spring and summer temperatures increase, pest insects can reach levels where they cause economic damage to crops earlier in the year. Warmer temperatures also can cause some insects to complete more generations in a year and therefore increase their population growth. Both effects from higher temperatures point to increased use of insecticides and pollinators near agricultural fields being more frequently exposed to these toxins. While pesticide exposure can kill colonies outright, I will discuss sublethal effects from pesticides, as these are far more common. Sublethal effects of pesticides in combination with other stress factors such as poor nutrition or Varroa can be the tipping point that causes colonies to be lost, especially during the winter or early spring.

Beekeepers face steep challenges moving forward and need local and national support to continue to provide the colonies needed to pollinate crops. I provide suggestions on how colony losses can be reduced. From a broader viewpoint, we all face challenges from climate change now and will even more so in the future. We quickly are reaching a point where everyone has in some way been affected by climate change. What will it take to reverse the rates of greenhouse gas emissions we produce and start a future where global warming declines? It will take a global effort, that is certain. The good news is that such efforts are underway. However, using alternative energy sources and adopting and developing green technologies are not occurring as they need to be throughout the world. In a survey conducted in the United States, more than a third of the people did not believe that the planet is warming or believe that global warming is caused by human activity.

As I tell the stories of plants, pollinators, and honey bee colonies, I have tried to convey my sense of wonder and deep reverence for the natural world and why we have a responsibility to care for it. I also emphasize connections within and among organisms and between our past decisions and present situation. Nature shows us that nothing happens in isolation and that there are both immediate and long-term consequences for our actions. While colony losses can be attributed to conditions in a given year, the continued difficulties in keeping colonies alive are due to past decisions about land use, energy sources, and overall lifestyles. Nature shows us that damage can be repaired, but it takes time and persistence and must occur at many levels. Reversing climate change will require cooperation and coordination of efforts in local communities and with those outside our borders, as we are connected by common hopes and needs. While this book is about climate change and honey bees, I hope it provides a broader perspective of how to move forward and that the natural world with its basis in interconnectedness and interdependence can provide the guidance we need.

Acknowledgments

I'd like to thank my family and friends for their unending support and for listening to my stories about honey bees, plants and flowers, forests, and wilderness. I also want to thank my advisors Clarence Collison and Roger Hoopingarner, all my colleagues in the bee research community, Robert Curry who co-created simulation models with me, and all those who apply mathematics to build models that generate deeper questions and understanding. I've learned so much from all of you. I also want to thank Jennie Durant for generously sharing her time and expertise, providing input on earlier versions of this book, and Emily Watkins deJong and Mona Chambers for sharing their photo albums and pictures seen throughout this book.

Introduction

I like to think in systems and see the connections among things. Once you start seeing connections that create systems, it is difficult to unsee them. This book is about the connections among honey bee colonies, plants, and the environment, and the effects of climate change on colony health. Some of the connections between climate change and colony losses are obvious, such as apiary destruction from floods or fires. Other connections are more subtle and have delayed effects on colony loss. The subtle impacts of climate change on colony growth and survival are emphasized in this book because they are revealed only by examining connections between honey bees and their environment as colonies move through time. For example, there are immediate effects of climate events such as summer heat waves or droughts on colony nutrition. There are also long-term repercussions that can extend into overwintering survival and colony growth the following spring.

There are three major themes woven through this book. The first is explaining systems thinking and how to shift from seeing separate individual parts to collective interconnected wholes. The second theme is applying systems thinking to honey bee colony biology, pollination, parasites and pathogens, and climate change itself. The final theme is that solutions to colony losses and climate change are available. I suggest colony management methods to reduce colony losses, but implementing the methods first requires consideration of the new realities we are facing.

Nature in general and honey bee colonies in particular lend themselves to systems thinking. By systems, I mean a set of interconnected components that together generate a specific product or function. Organisms are comprised of numerous systems. There is a digestive system, nervous system, and respiratory and circulatory systems that are constantly interacting and enabling the organism to function. In honey bees, there are also interacting systems among groups of bees in the colony. The interactions generate many essential colony functions, such as temperature control, nest defense, brood rearing, and food collection, distribution, and storage.

There are also connections among organisms and their environment that sustain populations and support individuals' health and reproduction. For example, flowering plants have connections with pollinators and rely on them for reproduction. From the plant's perspective, the relationship is built on attracting pollinators using colorful and aromatic floral displays and then rewarding the pollinators they attract with nectar and pollen. As pollinators collect rewards, they move pollen within and between flowers and initiate seed set. From the pollinators' standpoint, finding flowers that provide nectar and pollen is essential for survival and reproduction because many pollinator species, including honey bees, rely solely on nectar and pollen to fulfill all their dietary requirements.

Complex systems like those within an organism, or a honey bee colony, or a flowering plant are comprised of multiple smaller systems called subsystems. The relationships that develop between organisms are built on connections between their subsystems. For example, plants have subsystems to produce flowers with colors and odors that foraging bees detect through their visual and olfactory subsystems. In honey bees, large patches of rewarding flowers can activate parts of a colony's foraging subsystem that includes communicating the location of the flowers to nestmates. The product of the interactions between the plant and pollinator subsystems is a population of foraging honey bees moving pollen within and among flowers and initiating seed set.

For the most part, we are not taught to think in terms of systems. We often obtain information as factual events that result from isolated incidents. There is also a tendency to separate the individual components of complex systems and then study the components in detail. While there is value in understanding the individual parts of a system, that approach alone might not be sufficient for understanding the system as a whole. Some behaviors and outcomes of systems emerge only when seen in their entirety. A case in point might be the causes of honey bee colony losses.

Since the 2006–07 season, beekeepers have lost 20%–40% of their colonies each year. In the 2024–25 season, average losses exceeded 60%. Investigations into the causes of the losses invariably arrived at the conclusion of varroa mites, or viruses, with pesticide exposure detected as background stress. However, except for catastrophic events such as severe pesticide exposure, floods, or fire, the colony loss probably was a product of a series of events that could have been set in motion weeks or months ago.

So how do we shift our thinking from seeing events as isolated incidents to outcomes of systems? In Chapter 1, I provide an overview of systems. A point that is particularly important to keep in mind is that systems are dynamic. Energy flows through systems, causing component values to change along with the products of the systems. For example, the rate that pollination and seed set occur changes throughout bloom. The components of the system are the number of open flowers on a plant and the size of the pollinator population collecting food from the flowers. The number of open flowers changes throughout the day and over the period when the plant is in bloom. Changes in flower number affect the number of pollinators visiting them and the rate of pollination and seed set.

Another aspect of systems that I introduce in Chapter 1 but recurs throughout this book is the time it may take for the outcomes of systems to be fully realized. Some outcomes occur in a relatively short period of time. Pollination and seed set occur over the few weeks when plants are in bloom. Conversely, colony losses may take months to occur, such as when they are infected with viruses transmitted by varroa mites in the spring and summer and then are lost over the winter. The effects of burning fossil fuels on the earth's climate have taken more than a century to be realized, and during that time, it appeared that there were no negative repercussions on the environment.

A key feature of many systems is the formation of feedback loops. Throughout this book, I will continue to identify feedback loops first discussed in Chapter 1. A feedback loop develops when the amount of material in one component of a system affects and sometimes regulates the amount in the other. An example of a feedback loop that develops in honey bee colonies is queen egg laying, brood rearing, and colony size. The amount of brood that a colony can rear depends on the population of adult bees. If colonies are large, many of the eggs that queens lay will be reared into new adult worker bees. As the number of new workers increases, so does the amount of brood rearing, and consequently, the colony grows. In this scenario, the momentum created by the feedback loop between brood rearing and colony size will cause colony growth to move in a positive direction that will increase with each brood cycle. However, feedback loops also can generate negative momentum. If a colony has a small population because of diseases or poor nutrition, few new bees will be available to rear new adults. With fewer new adults, there will be fewer bees to rear brood, and the colony will be in a downward spiral to colony failure. This point about feedback loops driving colonies to either grow and flourish or be set on a course toward failure will come up throughout this book.

The impact that climate change is having on honey bees can be fully appreciated by examining the subsystems that comprise a honey bee colony and how they enable it to respond and adjust to the environment. For example, subsystems in colonies create stable temperatures in the hive, especially in the brood area in response to ambient temperatures. Plants are also comprised of subsystems that can detect environmental conditions and respond in ways that keep the plant alive. Perceiving and adjusting to changing environments is a fundamental feature of successful systems and is largely the product of the numerous interactions among subsystems. The range of conditions where systems can adjust and regain equilibrium are the system's boundaries.

Both honey bees and flowering plants have broad system boundaries, enabling them to establish populations under a wide range of conditions. Honey bees reside on every continent except Antarctica. Flowering plants that have established relationships with pollinators (i.e., angiosperms) comprise at least 90% of the plant species in most landscapes. Despite these wide system boundaries, honey bees are finding more areas that are inhospitable, and colony losses are occurring at unsustainable levels. Flowering plants also are experiencing levels of stress beyond their abilities to adjust and, in some cases, are unable to establish in habitats where they once thrived.

When conditions are outside the boundaries where systems can adjust, the stress makes the system vulnerable to a crash, especially if additional stress is exerted from outside the system. For example, climate change events such as high temperatures and lack of rainfall that limit plant growth and flowering can cause nutritional stress. The colony can adjust to nutritional stress by rearing less brood until pollen and nectar begin flowing again. However, if the colony also is experiencing parasitism by varroa mites and infections from the viruses they carry, the combined stress can create conditions beyond where the colony can adjust. The result is colony loss that looks like varroa mites, but the root cause was the stress caused by environmental conditions.

Throughout this book, I first provide a big picture view of the system I am explaining. For example, in the chapter about varroa mites, I begin by discussing parasites in general. This is followed by details about varroa mites that show that the mite employs basic parasite strategies that enable it to access their host (worker and drone larvae, pupae and adults), feed on it with little resistance, reproduce, and then disperse to other hives to find new hosts.

When discussing pollination, I begin with an overview of the fundamental biology of plants followed by a description of photosynthesis. The ability of plants to capture sunlight and use it to combine CO_2 and water to produce simple sugars and oxygen gas (i.e., photosynthesis) is the foundation for life on earth. Plants combine the simple sugars into complex molecules and assemble them into interacting sophisticated structures such as leaves, stems, roots, wood, seeds, pollen, and nectar. Appreciating how plants function and interact with their environment and with pollinators provides a basis for understanding how the stresses from climate change, such as excessive heat or drought, disrupt plant–pollinator relationships. The disruption can culminate in colony losses and prevent plants from setting seed and reproducing.

When discussing climate change, I begin with the details of the earth's carbon cycle and how this system maintains atmospheric carbon levels and in turn regulates air and ocean temperatures. The systems that recycle and sequester carbon are what enable life on earth to survive, evolve, and thrive. The carbon cycle has effectively kept atmospheric CO_2 levels at or below 300 parts per million for millennia despite volcanic eruptions and fires that release large amounts of greenhouse gases and particulates into the atmosphere. However, human activity over the past 200 years has taken the systems that regulate the release of CO_2 past the boundaries where atmospheric levels can remain at or below 300 ppm. Levels have been increasing since the 1950s, with the latest levels reaching 430 ppm (https://gml.noaa.gov/ccgg/trends/weekly.html?utm_source).

I provide a historical perspective to explain why levels of atmospheric CO_2 and other greenhouse gases have risen, especially in the 21st century. I begin with the period between 1760 and 1840, known as the First Industrial Revolution, and from the 1870s to the early 20th century (the Second Industrial Revolution). These periods are marked with rapid innovations in machines and transportation powered by fossil fuels. The demand for fossil fuels that began in the mid-19th century has not slowed but instead has continued to increase globally. In 2024, coal use reached an all-time high of 8.8 billion tons (1000 kg; 2204 lbs) (https://www.iea.org/energy-system/fossil-fuels/coal?). Oil and natural gas use also are at their highest levels since they began increasing in the 1950s (https://ourworldindata.org/energy-production-consumption?utm_source=chatgpt.com).

I provide this historical perspective of climate change for two reasons. The first is to illustrate the numerous systems built over nearly two centuries that rely on a steady supply of fossil fuels. Even though the effects of burning fossil fuels and releasing greenhouse gases into the atmosphere have been known since the mid-19th century, they were largely disregarded. Only relatively recently are we seeing the effects that burning fossil fuels have on weather patterns, ecosystems, and places where communities can be built and sustained. Agriculture and food security also are being affected, as is the beekeeping industry.

The second reason for providing a historical perspective is to capture the size of the challenge and effort required to develop and adopt new systems to replace fossil fuels. Systems built around the collection, production, and use of fossil fuels are embedded in our economies and lifestyles even though the long-term products of the system may make it more difficult for us and future generations to inhabit our planet. Global cooperation and commitment will be needed to reverse current trends in fossil fuel use. The good news is that many solutions are available to reduce carbon emissions. What is needed is the will to adopt them on a global scale.

Solutions are also available to beekeepers to reduce colony losses. The tried-and-true methods used to manage colonies no longer provide the assurance that colonies will thrive in our current environment. New management schemes for overwintering, varroa mite control, and colony nutrition are being developed for the current climate conditions. However, changing certain management strategies alone will not reduce colony losses. Beekeepers need commitments and cooperation from growers, land owners, environmental protection agencies, and local and national groups focused on preserving pollinators.

The final portion of this book contains my thoughts on how we might overcome the challenges of climate change. The solutions lie where the book begins, connections. Reducing fossil fuel use and reversing climate change requires cooperation on a vast scale and a rethinking of our connections with nature and with each other. How did Western civilization accept that we were separate entities from the natural world and perceive nature as either a threat or a resource to be harvested? Again, I take a historical perspective to address these questions, beginning with the earliest writings of philosophers and ancient religious texts that provided the framework for this thinking. Played out over time, the conscious or subconscious belief in our separateness generated an assumption that our assaults on nature and each other have no repercussions on our own health or survival.

The connections we have with each other and how we draw the circle around community also need to be reexamined if we are to address the current climate challenges. We live in communities that we construct, but often they are circumscribed and can lack diversity. Broadening our perception of community will be essential for dealing with the fallout of climate-related disasters that are displacing people around the globe. Whether we accept these individuals and fully integrate them will depend on how we define our communities and decide who can feel a sense of belonging there. Nature can provide a blueprint for how to integrate new individuals as natural communities thrive on diversity and multiple connections among different groups.

Some final thoughts in this book are directed at our responsibility as individuals in addressing climate change. We know that the changes in lifestyles that we make as individuals will have little effect on global carbon emissions. However, there is power in collective action. Just as in the past, where large social movements began with the actions of small groups, so can our decisions to reduce fossil fuel use gain the momentum needed to reverse climate change. Ideas and movements spread faster now than ever before because of globalization and ease of communication, travel, and commerce, so there is reason for optimism that change can happen quickly.

The motivation to adopt new forms of energy can arise from the impacts of fossil fuels, especially at the local level. These include destruction of ecosystems, air and water pollution from refineries and increasing frequency of extreme weather events such as hurricanes, tornadoes, summer heat waves, floods, and fires. The impacts are affecting communities without regard to location or socioeconomic status. We are rapidly reaching the point where many people have either been affected by climate events or know someone that has.

I conclude this book by suggesting some starting places for the changes in society that will be needed to reverse climate change. These involve expanding our imaginations and rethinking our place in the natural world. Nature can provide a starting place for understanding basic tenets of successful communities, as does learning about honey bee colonies. Honey bees are successful for many reasons, not the least of which is their ability to take a long view. The activity of a colony in the spring is to ensure its survival in the present time and months later during the winter. The behaviors of individual adult bees are for the immediate benefit of the colony but also to benefit those that will comprise the colony in the future. The same long view is needed to change our behaviors that are contributing to the climate crisis and set a course to secure the future for those who follow.

Chapter 1

All outcomes are products of systems

Everyone has had the experience where once we connect with a new product, image, or concept, we tend to see it everywhere. The experience we share is called frequency bias or the technical name, the Baader–Meinhof frequency illusion. If you buy a certain make or color of car, you start seeing them everywhere. The number of cars like the one we just bought did not suddenly increase. We just notice the cars more often. The goal of this chapter is to introduce you to systems thinking so that you start seeing systems everywhere, particularly in nature and in the biology and behavior of honey bee colonies.

A system is defined as a set of objects or components (e.g., organisms, cells, and molecules) that are interconnected or interdependent so that they generate behaviors and patterns over time [1]. To recognize a system, you need to see connections between and among objects or events that produce an outcome. We often think of outcomes as isolated incidents. Things happen because of current circumstances. However, when you start seeing systems, you know that there are no isolated incidents. In fact, nothing is really isolated. When an event occurs, it is the product of interactions among components in a system. The event or outcome might have been set in motion by a series of events in the recent or distant past. For example, the size of a honey bee colony at a given time is the product of recent conditions such as the availability of flowering plants or presence of pathogens in the colony. Present conditions also are the product of past decisions such as swarming or queen replacement. When we learn to see relationships between objects and outcomes, we begin to understand how systems work and that the outcomes we see are products of the structure that generated them. We will come back to this point often throughout this book, especially as it relates to the dynamics of honey bee colonies and climate change. In the case of climate change, systems that were put in motion to fuel innovations and lifestyles more than 200 years ago have created the situation where we now find ourselves. The good news is that solutions that change system behaviors often are within the parts of the system itself.

Complex systems are all around us. They keep traffic flowing, production lines moving and store shelves stocked. Anywhere there is a process, there is a

The Honey Bee and Climate Change. DOI: https://doi.org/10.1016/B978-0-443-28995-8.00003-9

system. Although systems designed by humans may be complex and even efficient, they do not compare with the elegance and intricacy of systems comprising the natural world. For example, honey bee colonies grow and survive because of the integration of systems at cellular, microbial, organismal, and behavioral levels that are coordinated among tens of thousands of bees. Flowering in plants and their attractiveness to pollinators is the integration of systems across landscapes, species, and ecosystems whose product is pollination and seed set.

Looking at nature through the lens of interconnected organisms can be overwhelming, but the more we learn about nature, the more we see these connections. Perhaps seeing connections among parts of nature requires reading and exploring broadly across ecosystems and seasons. This was the case for the 18th century German scientist and geographer Alexander von Humboldt, who first proposed "to represent nature as one great whole, moved and animated by internal forces." In his five-volume work, *Kosmos* published from 1845 to 1862, von Humboldt pioneered the concept of the "unity of nature." He proposed that "the natural world is a web of intricately interconnected dynamic elements constantly in dialog with each other" [2]. From von Humboldt's perspective, nothing lives in isolation, but instead every organism, whether deep in the soil or on the tops of trees, is interconnected. The connections are at micro- and macroscopic levels and enable species to respond to each other and to their environment.

Considering nature as an interconnected web was a departure from the reductive approach of von Humboldt's contemporaries. From von Humboldt's time to the present, the reductive approach deconstructs individual parts of processes and examines them in detail. Individuals become specialists in a particular component of a process. An underlying assumption of a reductionist approach is that organisms can be understood by studying them with a series of increasingly narrow focus (e.g., groups → individuals → cells → genes → molecules → atoms → elementary particles) [3]. A reductionist framework is certainly a useful way to understand complex processes because breaking them down into individual parts makes them simpler and tractable [4]. However, processes studied in isolation and at the level of their smallest parts ultimately need to be put back into the bigger system where they operate to completely understand their behaviors and limitations. In the end, we have to reconnect the dots.

The natural systems we see today have been adapting to changing conditions for more than 3.5 billion years. Successful strategies are passed on to offspring and repeated across species. For example, the same biochemical systems to extract energy from food occur in numerous unrelated species including humans and honey bees. Strategies to enhance interactions among species also have evolved so that "lines of communication" are established to produce specific outcomes. A good example is the pollination of flowering plants by honey bees. Plants have systems that produce flowers and subsystems that synthesize pigments for flower color and volatile chemicals that emanate odors. The flower odors and colors have evolved so that they can be detected by honey bees and other pollinators. Nectar and pollen are produced by additional subsystems in flowers and these resources provide for the nutritional needs of pollinators. Pollinators detect the colors and odors from the flowers and associate them with food rewards (i.e., nectar and pollen). While bees collect the nectar and pollen, they pollinate the flower and initiate seed set.

When looking at the big picture of interconnections among organisms and their environment, we often see repeating patterns and loops that create cycles. Cycles are present throughout nature and often are driven by the seasons. Cycles also occur within individuals in cellular division, physiology, and biochemistry. Recognizing cyclical aspects of systems requires cyclical rather than linear thinking. With linear thinking, we use logical, step-by-step, cause-and-effect perspectives to understand a process. Linear thinking is useful in mathematics when solving equations and word problems or when assembling complex structures or machines. With linear thinking, one step follows the next.

Linear thinking also can be highly effective when conceptualizing how components in a system might fit together. However, understanding dynamic systems and recognizing the cycles and repeating patterns they generate requires moving past linear and into cyclical thinking. We will apply cyclical thinking to discuss the dynamics of honey bee colonies throughout the year. We also will discuss the carbon cycle and the ways carbon is sequestered and then released into the atmosphere and bodies of water in a cyclical manner. It is the cycling of carbon through air, water, rocks and soil, and living organisms that makes life on earth possible. One last thing about cycles; like any system, factors that affect any one component of the cycle will impact the outcome of the entire cycle.

So, how do you recognize a system? How will you know when you see one? Perhaps the most obvious quality of systems is that they are dynamic. By this I mean that the products of systems are constantly changing over time especially as they interact with changing environments. For example, a honey bee colony has a different number of bees in it each day because of the amount of pollen and nectar bees are collecting and the amount of brood that are reared. Systems also can be recognized by their basic parts: individual components, interconnections and flow of resources or information between and among the components, and an end product. The end product of a system is generated from the interconnections among components. Measuring a product at a particular point in time reveals the "system state." Measuring products over time and under different conditions reveals the "system behavior" [1]. Looking at a honey bee colony in a systems framework, we see that two components, workers and queens, are connected in that workers care for the queen. The product of the connection is eggs laid by the queen that can develop into new bees (Fig. 1.1).

The number of eggs laid by the queen at any point in time reveals the system state. If we measure egg laying and the number of adult worker bees over a period of time, and see that they increase, we can conclude that the system behavior is one of colony growth.

Although the outcomes of some system might be obvious, in many cases, unexpected outcomes emerge as synergisms among components make systems more than the sum of their parts. An example of an outcome that might be readily apparent is when a colony finds a rich nectar source and activates its foraging and food storage systems. Workers are recruited to the flowers, and within a week or so, the bees have collected combs full of honey and pollen. Outcomes from other systems however, might not be so obvious and may take many years to manifest.

The effects of climate change that we are now experiencing are the product of systems set into motion more than 150 years ago. The increasing release of carbon in the form of CO_2 and greenhouse gases from burning fossil fuels accumulated in the atmosphere and oceans over many decades. The levels of carbon cause air and water temperatures to rise, and these affect weather patterns, sea levels, and ultimately ecosystem composition and frequencies of fires and floods.

The terms and concepts explained in this chapter will be revisited in subsequent chapters to explain how interactions among honey bees in a colony and between foraging bees and flowering plants generate behaviors and products that affect colony growth and survival. Climate change also will be described in a systems framework because of the connections among increasing greenhouse gas levels, rising air and ocean temperatures, and highly destructive weather events. The effects of climate change on honey bees and flowering plants also will be described in a systems context because they

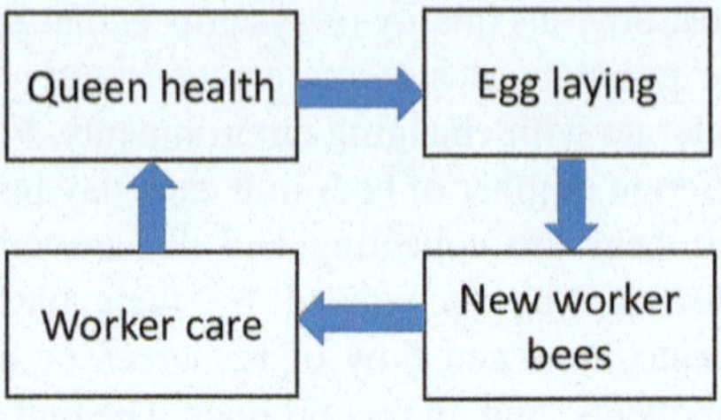

FIGURE 1.1 Connections among worker care for the queen, her egg laying, and the number of new worker bees produced that can care for the queen. The system generates egg laying and the production of new worker bees from the connection between workers caring for the queen and queen health.
Diagram by Dr. Gloria DeGrandi-Hoffman.

show fundamental attributes of systems behavior. Specifically, stress on one part of a system reverberates through the entire system, with repercussions being felt immediately or sometime in the future.

1.1 Feedback loops and system momentum

An integral part of systems regardless of the type is the feedback loop. A feedback loop is established between two components when the amount of material in one affects the amount in the other (Fig. 1.2).

There are two types of feedback loops. The first is called a "reinforcing or positive feedback loop." The relationship between components in a reinforcing feedback loop generates products that reinforce their own production. Reinforcing feedback loops create momentum that courses in a positive or negative direction and can cause exponential growth followed by crashes. Whichever direction production is moving in a reinforcing feedback loop creates momentum in that direction and makes it increasingly difficult to reverse or to reestablish equilibrium [1].

Let me clarify with an example. There is a relationship between the size of the adult worker population in a colony and the amount of brood that can be reared. When honey bee colonies have a healthy queen, plenty of food, and warm sunny weather so bees can forage, the colony will rear large amounts of brood. The brood develops and many new adult workers emerge and join the adult population. As numbers of new workers increase, more brood is reared and the feedback loop between workers and brood rearing has positive momentum and the colony grows. This scenario occurs in honey bee colonies in the spring. The feedback loop between workers and brood also can have negative momentum. For example, if a colony has a small adult population, brood rearing will be limited by the number of adult workers. If we add a

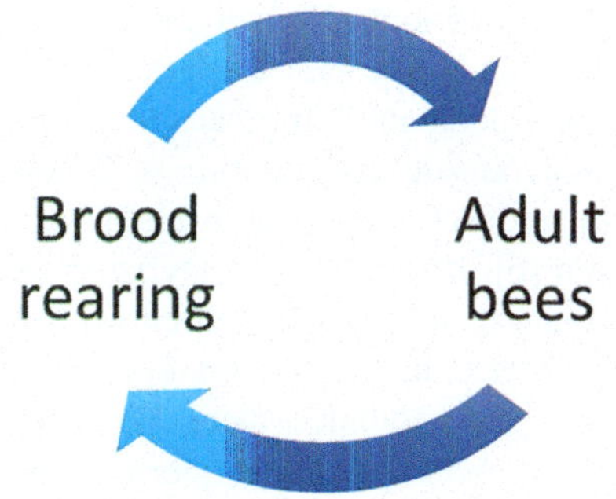

FIGURE 1.2 The feedback loop between brood rearing and adult bees. More adult bees enable more brood to be reared, producing more adult bees resulting in colony growth.
Diagram by Dr. Gloria DeGrandi-Hoffman.

mortality factor so that workers are dying faster than new ones are emerging, the population declines. If there are fewer workers to rear brood, less brood will be reared. Now the feedback loop picks up momentum in a negative direction. There are fewer bees to rear brood so less brood are reared, so there are fewer bees, and momentum is spiraling downward [5]. The momentum generating a downward spiral can be reversed but it must be addressed early in the cycle. Once momentum in a negative direction is too great and cannot be reversed, the colony will die or become so weak that it is robbed by bees from another hive.

A second type of feedback loop is a "balancing feedback loop." Systems with balancing feedback loops have elements that constrain the growth of any one product and avoid boom or bust growth. In nature, a balancing feedback loop can contain factors that can constrain population growth such as low reproductive rates, or a high tendency to migrate or disperse. Parasites that do not harm their hosts often have balancing feedback loops that regulate their impact and allow the parasite to coexist with their host.

Before we move on from the concept of feedback loops, let me tell you a few more things about them. The effects that feedback loops have on a system may not be immediate. Going back to our example of a downward spiral in a colony population, it might take several days or weeks for the colony to decline and collapse, especially if the colony is large. Colony losses over the winter and spring are examples of delayed effects set in motion by the worker-brood feedback loop. If factors such as Varroa mite parasitism and virus infections are high in the fall, the lives of the overwintering bees can be shortened. When foraging resumes in the spring, the short-lived bees may die at rates that exceed those that new bees emerge. Each day there will be fewer bees to rear brood causing the downward population spiral. Although the colony is lost in the spring, the conditions that culminated in the loss were initiated through the brood-worker feedback loop established the previous fall.

We will return to feedback loops throughout this book because the subtle effects of climate change on honey bees often occur through feedback loops. That brings me to the next point about feedback loops; they can be affected by factors from inside and outside the system. A factor affecting the brood rearing–colony growth feedback loop from inside the system could be the age and health of the queen [6]. If the queen is laying progressively fewer eggs, there will be fewer adults to rear brood, setting off the downward spiral of population decline. To reverse the momentum, systems need to be activated in a timely manner. In honey bee colonies, there is a queen replacement system that is activated when workers detect the queen is failing. The sooner the detection occurs, the better the chances of reversing the colony decline before it reaches the point of no return.

Adjusting to factors inside a system that affect feedback loops can be somewhat straightforward; responding to external factors can be much more difficult. Honey bee colonies are continuously interacting with the environment, particularly when it comes to food. When flowering plants are available and pollen and nectar are flowing into the colony, the feedback loop between adult workers and the queen, can drive brood rearing and population growth in an ever-accelerating upward direction. However, when bloom is waning and less food is available, brood rearing declines. Unless new food sources are found, the decline in brood rearing can shift the momentum of the feedback loop in a negative direction. When less brood is reared, there are fewer new adult bees to rear brood so the feedback loop momentum is coursing in a negative direction ultimately causing an ever decreasing colony population.

1.2 Qualities of functional systems

Highly functional systems have safeguards against internal factors that can cause crashes. Functional systems also can adapt to a wide range of external conditions. Trees in old growth forests are a great example of highly functional systems. Trees in those forests have lived for hundreds and sometimes thousands of years and experienced and adjusted to wide ranges of environmental conditions and disturbances. Such system longevity is due to three general properties: resilience, self-organization, and formation of hierarchies [1]. These properties enable systems to adapt to changing environments and conditions. Functional systems also have developed multiple subsystems existing at different places in a hierarchy. The subsystems work together with the product of one subsystem feeding into or controlling another.

Systems with built-in resilience have the capacity to withstand and recover from a wide range of challenges because they have built-in flexibility and can spring back into shape after sustaining pressure or stress. The resilience comes from timely recognition of the stress, and a modulated (rather than all or nothing) response. An example of a resilient system is the one honey bees activate to maintain internal hive temperatures within the narrow range needed to rear brood and remain active.

If we look at the thermoregulation system in honey bee colonies, we first see flexibility in how workers respond to heat and cold. When temperatures are high, foragers collect water and nestmates use it for evaporative cooling in the hive. Workers aslo spread out across the combs and gather outside the hive entrance to reduce heat inside the hive. If temperatures are low, bees huddle tightly together, vibrate their wing muscles and generate heat to keep the colony warm [7]. Rapid response to ambient temperatures can be the difference between a colony dying from overheating or from the cold. Although the response is rapid, it also is modulated. Rather than the entire colony responding, groups of workers largely determined by their genetics, readily detect disturbances and initiate appropriate action.

This brings me to a last point about functional systems; and that is "boom or bust" behaviors. Whenever the amount of product from a system rapidly increases, more resources will be needed to sustain it. Without increasing resources, production rates decline earlier and faster than if the flow of resources and accompanying growth occur at an even keel. Revved up reinforcing feedback loops that cause sharp spikes in growth occur in both natural and man-made systems, and they almost always end badly. Rapid growth followed by a precipitous decline in a system's product often is referred to as a "crash." This behavior can occur in honey bee colonies that rapidly build populations in spring when trees are flowering and nectar and pollen is abundant. Large populations need large amounts of food to sustain them. If there are no flowers because of high heat or drought, or the weather turns cold and bees cannot forage, nectar and pollen stores will be used quickly and populations can decline sharply unless new food sources are found.

1.3 Self-organization

Once you start noticing that systems are everywhere, you also may see that the ones that generate complex processes are made up of many smaller subsystems [1]. Our bodies have numerous subsystems that participate in digestion, breathing, circulation, and more. Honey bee colonies also have numerous subsystems that are activated when colonies need to care for brood, forage for food, build comb, and store honey. A major difference between the subsystems in our body and those of a honey bee colony is that many of our subsystems are under a central control (i.e., the brain). The central controller perceives a stimulus and then relays the information to peripheral subsystems to generate a response. In honey bee colonies, there does not seem to be a central control site for processing incoming information. Instead, groups of workers perceive a stimulus and generate a response. The response begins with interactions among bees at a local level (i.e., a particular location in the hive) but can spread across the hive and ultimately involve hundreds and sometimes thousands of bees. The interactions among individuals that produce behaviors based on local information and rules yet ultimately generate global patterns or responses is a process called self-organization [8,9]. Complex colony-wide processes such as thermoregulation, brood care, foraging, and comb building occur through self-organization.

Honey bee colonies have numerous self-organized subsystems that are coordinated to generate processes that are critical for colony survival. In our simple model of colony growth, the queen lays eggs and adult bees rear the brood into new adult bees. The brood rearing system requires nectar and pollen. To collect nectar and pollen efficiently while still rearing brood, colonies divide the tasks of foraging and brood rearing between different groups of bees. Workers that are foraging interact and generate behaviors such as which flowers to forage based on local information provided by other foragers. Food flows into the colony through the foraging subsystem and is available to the nurse bees and larvae that are interacting in the brood rearing subsystem. New adults are produced and join the brood rearing subsystem. After several weeks, the workers enter the foraging subsystem and continue the flow of food into the brood rearing subsystem and the colony at large (Fig. 1.3). When resources become limited, brood rearing slows down and may eventually stop until resources again become available.

Coordination among subsystems helps to build resilience and efficiency into systems. Relationships among components within a subsystem are stronger (i.e., have greater effects on each other) than between subsystems, and this enables subsystems to respond to changing conditions with greater immediacy. Let me give you an example to clarify these aspects of subsystems. Picture foraging bees bringing nectar into a colony. The nectar is passed to a nestmate near the colony entrance, and from there, it is passed among other bees in the colony. Excess nectar is stored in wax comb and concentrated into honey. During abundant nectar flows, incoming nectar can overrun the capacity of the bees to store it. When there is no longer open space on combs and individual workers have reached their capacity to hold nectar in their honey stomachs, workers will no longer accept nectar loads from incoming foragers. When these conditions occur, a physiological subsystem is activated enabling worker bees to convert sugars from nectar and honey into wax. Foraging will stop and workers in the wax-producing subsystem will connect with nestmates in the comb-building subsystem to build wax cells and create

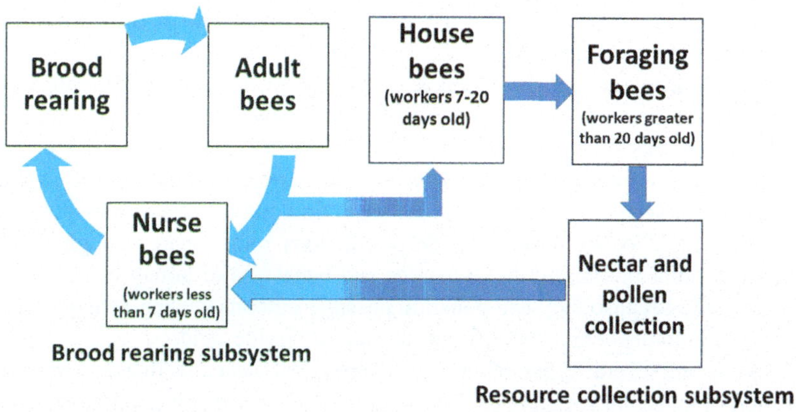

FIGURE 1.3 Example of subsystem coordination in a honey bee colony. Brood rearing produces new adult bees that become nurse bees that rear brood and continue the cycle that produces new bees. The "brood rearing subsystem" requires young worker bees and nectar and pollen to function. The "resource collection subsystem" consists of worker bees that are older than 21 days and have converted into foragers. The foragers collect nectar and pollen and make the resources available for brood rearing.
Diagram by Dr. Gloria DeGrandi-Hoffman.

additional storage space. When there is room again to store food, workers will resume accepting nectar brought into the colony by foragers and foraging will resume [10].

The interconnectedness among subsystems is sometimes referred to as a hierarchy [1], but I like to think of it as a network. Information and products are exchanged through various points of intersection (nodes) positioned between different subsystems. For critical functions, the nodes are positioned a short distance from feedback loops, so the system can respond to disruptions more quickly. In our example of foraging and comb-building subsystems, the node for the two subsystems is at the colony entrance. When foragers return from the field and cannot find nestmates to receive their nectar loads, activity in the "resource-gathering" subsystem is suspended, and the "comb-building" subsystem is activated. Notice how each of the subsystems is self-organized and operating on local information and interactions among groups of workers that when coordinated generate a colony-wide response.

1.4 Models, boundaries, and unexpected consequences

We can think about systems in an abstract way and imagine how components might connect. Diagrams of system components with arrows connecting them (i.e., flow diagrams) are useful for visualizing how a process might work. Whether our understanding about a system is correct can be tested by building a model of the system. Models are representations of what we think we know about a system and how the components might interact with each other and respond to the environment.

Models that capture the essence of a system and accurately reflect its behavior are powerful tools, especially if the model can predict outcomes under conditions that may be difficult or dangerous to test in real life. For example, NASA uses models to test what could happen to astronauts or a space craft under given sets of conditions. Better to find out where and how possible disasters might occur through model predictions rather than through actual events. Models predicting the dynamics and behaviors of honey bee colonies are valuable because they can generate possible outcomes when actual studies with live colonies would be difficult to conduct in the field.

There are several types of models. A physical structure that is a smaller version of an actual structure can be a model and be used to examine structural stability and limitations under a range of conditions. Often models are computer programs containing equations that mathematically describe the actual system. In all cases, the model is a simplified version of the actual system and generates predictions about its behavior. We can test how well our model captures reality by exposing it to different conditions, and when possible, comparing the predictions with actual outcomes. Each outcome is a snapshot of the model's performance. Putting together many snapshots generated under a range of conditions reveals trends that uncover the model's behavior. How closely the behavior of the model resembles the actual system's behavior indicates whether we have identified the important components of the system and correctly described their interconnections.

First attempts at building models can be difficult. Trying to describe in specific terms what we think we understand often reveals gaps in our knowledge and places where connections might be missing. The first step in building a model is to clearly define its product. For example, the product of a model of a honey bee colony might be the size of the population over some period of time. The next step is to determine all the factors that affect the product of the model.

A useful technique during the early stages of developing a model is one that originated in physics and engineering called "free body modeling." Free body modeling is essentially a brain storming session where all the factors that could affect the product of the model are drawn as arrows that represent forces (Fig. 1.4). In the case of a honey bee colony model, factors that exert forces on population size could be the queen's age and her egg-laying rate, weather conditions, amount of pollen and nectar, pathogens, parasites, and pesticides. All of the factors we identify during free body modeling might not be included in early versions of the model, but it is useful to be aware of them as they might need to be added later.

The next step in building the model is to connect the factors from the free body model to produce an image (flow diagram) of how they might fit together. In the case of our honey bee colony model, the connections among the factors create a cyclical structure. The colony population size at any time is

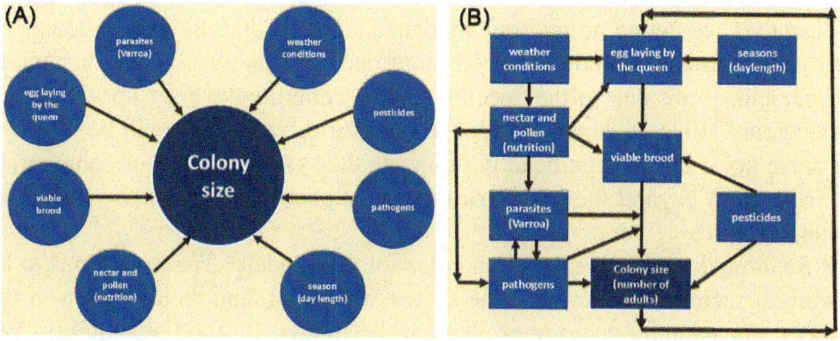

FIGURE 1.4 The two steps in developing a model to predict the size of a honey bee colony population through time. Step 1 is to create a free body model (A) that includes all the factors that can exert forces that shape the size of the honey bee colony population. Step 2 is to connect the factors into a flow diagram of the model (B) to create a visual image of how the factors might interact and affect colony population size through time.
(A) Free body model diagram by Dr. Gloria DeGrandi-Hoffman. (B) Flow diagram by Dr. Gloria DeGrandi-Hoffman.

not just the product of the model but a factor that affects the number of eggs that are reared to adults. Many models of biological systems have a cyclical structure.

After conceptualizing the model, the next step is deciding the type of model to build. Models that predict population sizes over time often are mathematical models. If a mathematical model is being constructed, equations need to be derived describing the interactions among factors. For our colony model, equations could be derived using data of population measurements collected from actual colonies throughout the year to estimate the effects of temperature or daylength on colony size. To estimate the effects of nutrition, differences in population growth could be measured in colonies having different amounts of pollen and nectar. Eggs laid by queens of different ages measured under the same weather and colony conditions could provide information on the effects of queen age on egg laying and colony growth. The process of formulating a model can generate numerous hypotheses and experiments that not only produce valuable data on factors that affect colony growth but also information that can be interpreted within the big picture of colony population dynamics.

One of the challenges of building models is to avoid making them too simple or too complex. If we take a holistic view of nature, it is safe to say that everything is interconnected through subsystems at the microscopic and macroscopic levels. However, building models of natural systems that are not overwhelmingly complex requires drawing boundaries. Establishing boundaries when describing a system and more importantly, when trying to predict system behaviors, takes a lot of thought. Some of the factors identified during the free boding modeling might be placed outside the boundaries at least in the beginning of developing the model. We need to be aware of boundaries we define around our models and know that they are products of our conscious and sometimes unconscious decisions and biases. The behaviors of our model are due to the specific sets of conditions we set up within the boundaries. What we leave outside the boundaries very likely will be revealed at some point if they significantly influence the system product or behavior. It is from areas beyond the boundaries that unexpected behaviors and outcomes often arise.

So, how do we define the boundaries for our models? The first step is to be aware of factors that can limit the system's product, and include them in the model. For example, colony growth can be limited by food availability. We can construct a model where queen egg laying, brood rearing, and colony growth are predicted based on foraging activity and amounts of nectar and pollen entering and being stored in the hive. To test if our model's predictions of egg, brood, and adult bee populations accurately capture actual population sizes, we can compare actual foraging activity, food stores, and population growth in colonies with values predicted by our model under given sets of conditions (e.g., restricting foraging or limiting food stores). How accurately our predictions resemble actual colony growth can reveal if factors other than foraging and food stores are affecting colony growth. If our predictions overestimate how rapidly the colony is growing, there probably are limiting factors that are not included in the model. These might be the rate of egg laying by the queen as affected by her age and health. Diseases that are causing brood or adult bee mortality also might be affecting colony growth. Measuring individual components of the model such as the number of eggs laid by the queen each day or the proportion that develop into adult bees can indicate if our model needs to incorporate new components to improve predictions.

Another way to gain insights from building models is to remove all limiting factors to determine potential growth of a product. Setting up a model that does not have limiting factors can reveal more nuanced effects on product growth. For example, we can set up our colony population model to have unlimited space and food and predict colony growth based only on weather conditions and daylength (i.e., season of the year). Predicted colony growth can be compared with growth measured in actual colonies that were provided with plenty of nectar, honey, pollen, and frames of empty comb. If our actual colony is not growing at rates predicted by the model, it could be because the queen is laying fewer eggs than expected or the lifespan of workers is shorter than we predict. We can set up experiments to test each of these possibilities. In this way, building models is a good way to direct research efforts and decision making. For example, if we find that colony growth is being limited by the lifespan of adult worker bees, we can test for pathogens or pesticides. If we are feeding colonies supplemental diets instead of pollen, and actual colony growth is lower than model predictions, there might be nutritional shortcomings in the diet that are causing malnutrition even if the supply of food is unlimited and bees are consuming the diet.

A final word about building and testing models; to identify their limits, models should be tested under a broad range of conditions and over extended periods of time. Predictions might be in line with reality under certain sets of conditions or over a given time interval. However, predictions might diverge from reality under different conditions or after some period of time. Some of the more subtle limitations of a model might be lying outside the boundaries we have established. Remember, the model we build is our best approximation of the system. A model is only as good as its ability to predict behaviors under new sets of conditions. In this way, models are built to be broken or at least constantly modified as our understanding grows. If predictions diverge from reality, something about how we understand and defined the system and the boundaries we built need to be adjusted. Unforeseen outcomes that challenge preconceived notions and take our thinking in completely new directions are an enormous benefit of systems thinking and model building. The types of insights that can be gained by considering an entire system often are not revealed by even the deepest examinations of individual components studied in isolation. In the following chapters on pollination, parasitic Varroa mites, and the annual colony cycle, I will describe how models predicted outcomes that differed greatly from those occurring in the field. The discrepancies led to new hypotheses and experiments that generated fresh perspectives on long held assumptions.

1.5 Conclusions and final thoughts

Aspects of systems thinking and model building such as feedback loops and subsystems will come up throughout the book beginning with a systems-based description of honey bee colonies. From there, I will describe pollination and seed set systems and show how they are connected to honey bees in ways that are mutually beneficial and sustaining. Threats to the survival of honey bee colonies such as Varroa mites also are described by the interconnections between mite and colony populations. Another concept that will be revisited time and again is that all outcomes are products of a system. Climate change is no exception. How we got to where we are today is a story of interlocking subsystems powered by carbon-based fuels that generate two main products: our current lifestyle and climate warming from greenhouse gases. The original intent of burning fossil fuels was not to release excess carbon into the atmosphere and oceans. However, taking the long view through thinking in systems and building models, it would have become apparent that reliance on fossil fuels over time was not sustainable or compatible with life on Earth.

The warming of our planet is generating weather systems that strain the resilience of plants, honey bees, and other pollinators and push them beyond where their system can adjust and restore growth, reproduction, and survival. Although climate change is affecting pollinators in general, managed bees and commercial beekeeping are especially impacted. Beekeeping practices that take the effects of climate change into account can mitigate some of the limitations, but it will be increasingly difficult to keep colonies alive and costly to do so. Reducing greenhouse gases and removing them from the atmosphere are the great challenge of our lifetime and will require modifying current systems and constructing new ones. Outcomes are the products of systems whether we consciously build the system or not. We can choose the outcomes we want through the designs of the systems we build.

References

[1] Meadows DH, Wright D, editors. Thinking in systems. London, UK: Earthscan; 2008.

[2] von Humboldt A. Cosmos: a survey of the general physical history of the universe. Meteorology in Nineteenth-Century Society. Routledge; 2025.

[3] Andersen H. The history of reductionism versus holistic approaches to scientific research. Endeavour 2001;25(4):153–6.

[4] Balanay R, Halog A. A review of reductionist versus systems perspectives towards 'doing the right strategies right' for circular economy implementation. Systems 2021;9(2):38.

[5] Betti MI, Wahl LM, Zamir M. Age structure is critical to the population dynamics and survival of honeybee colonies. R Soc Open Sci 2016;3(11).

[6] Tarpy DR, Gilley DC, Seeley TD. Levels of selection in a social insect: a review of conflict and cooperation during honey bee (Apis mellifera) queen replacement. Behav Ecol Sociobiol 2004;55:513–23.
[7] Fahrenholz L, Lamprecht I, Schricker B. Thermal investigations of a honey bee colony: thermoregulation of the hive during summer and winter and heat production of members of different bee castes. J Comp Physiol B 1989;159(5):551–60.
[8] Levin SA. Self-organization and the emergence of complexity in ecological systems. BioScience 2005;55(12):1075–9.
[9] Saetzler K, Sonnenschein C, Soto AM. Systems biology beyond networks: generating order from disorder through self-organization. Semin Cancer Biol 2011;21(3):165–74. Available from 10.1016/j.semcancer.2011.04.004.
[10] Pratt SC. Collective control of the timing and type of comb construction by honey bees (Apis mellifera). Apidologie 2004;35:193–205.

Chapter 2

Honey bee colonies from a system science perspective

2.1 Introduction

Prepare to be amazed by the intricate workings of a honey bee colony. I will begin with the basics of honey bee biology and then describe the coordination among worker bees performing diverse behaviors enable a colony to respond to ever changing conditions inside and outside the hive. The interactions among groups of worker bees performing similar tasks can be seen as subsystems. Brood rearing, food gathering, nest building, maintenance, and temperature control are subsystems that are activated when conditions and needs arise. The coordination of subsystems ensures overall system (i.e., the colony) survival.

The subsystems that comprise a colony of bees are flexible and responsive to changing environmental conditions. They exist in the colony at large, and also within individual bees. The physiology and neurobiology of bees are internal subsystems that generate development, immunity, food processing, and the diverse and complex behaviors performed by workers throughout their adult lives. The behaviors are coordinated among bees through chemical signals and interconnected flows of resources and information.

A final word about honey bee colonies; their survival is based not just on numbers of workers but also on their ages. Throughout this chapter and in others, I will refer to a colony's age structure. By age structure I mean the number of workers of different ages (age is expressed as the number of days the worker has been an adult) that are in the colony population. Colonies with large numbers of older bees, with perhaps only a few days to live, will cause populations to decline if the death of older workers is not offset by the emergence of new workers. In the spring, colonies can have an age structure with large numbers of young bees and few old bees especially if the colony began rearing brood before weather was suitable for foraging. New workers will emerge and accumulate in the hive and older workers reared the previous fall will die shortly after foraging resumes. The age structure shifts in the fall when brood rearing slows or stops and workers continue to age. Preserving

The Honey Bee and Climate Change. DOI: https://doi.org/10.1016/B978-0-443-28995-8.00002-7

young long-lived workers in the fall and over the winter is a key component for colony growth and survival in the spring.

2.2 Brood rearing and the annual colony cycle

Generally speaking, honey bee colonies have a single queen, thousands of worker bees, and depending upon the time of year, hundreds of drones (males) (Fig. 2.1). The queen is the only one that lays eggs, so the colony is composed of her offspring. When the queen lays eggs that are fertilized by sperm, they develop into females (worker bees). Unfertilized eggs become drones (males). Although worker bees are females, they do not lay eggs when a queen is present. Chemical compounds produced by the queen and worker larvae keep worker ovaries from developing and prevent them from producing eggs.

The queen communicates her presence and egg-laying capacity primarily through chemical cues (i.e., pheromones). For example, queens produce a pheromone (queen mandibular pheromone—QMP) that is present on the outside of their body [1,2]. QMP is a mixture of compounds made by the queen and released by the mandibular glands situated on the sides of her head. The amounts of each compound changes before and after the queen mates. When she is a virgin, the components of QMP attract drones during mating flights. After mating and establishing herself in the colony, the formulation of QMP and other compounds produced in glands on the queen's body attracts workers

FIGURE 2.1 **Honey bee workers, drone, and queen.** Worker, drone, and queen honey bees on comb in a hive. The drone is larger than the workers and has contiguous eyes that differ from workers that have a space between the eyes. The queen is marked with a dot on her thorax so she can be easily found in the hive. The workers surrounding the queen are her retinue. The workers are attracted to the queen and form a retinue because she is releasing a blend of chemicals called queen mandibular pheromone.
Photographs by Emily Watkins DeJong (workers and drone) and Mona Chambers (queen).

that form a retinue around the queen [3]. Some workers in the retinue pick up the compounds through physical contact and transfer them to nestmates throughout the hive. QMP and other compound produced by the queen are associated with many other activities in the hive such as stimulating construction of worker cells and inhibition of drone and queen cell construction (see refs. [3]).

QMP is detected by workers through physical contact with the queen and spread throughout the hive through bee-to-bee contacts. In addition to "contact pheromones" like QMP, there also are volatile compounds such as E-β-ocimene that signal a laying queen [3,4]. When a queen is failing, changes in her chemical cues signal workers to activate the queen-rearing subsystem to produce a new queen.

The brood (eggs and larvae) produced from the queen's eggs are fed and cared for by young worker bees (nurse bees). The nurse bees feed the larvae secretions from glands in their head called hypopharyngeal glands (HPGs) (Fig. 2.2). HPGs are located in front of the brain and between the compound eyes. HPGs start developing while workers are still in the pupal stage about a week before emergence as an adult. By day 3 as an adult, the HPGs can secrete royal jelly. The size and structure of the HPGs change throughout the life of a worker depending on their age, diet, and activity. In nurse bees, the size of HPGs is determined by the amount of pollen they consume and its nutritional content. The more pollen nurse bees consume, the larger their HPGs.

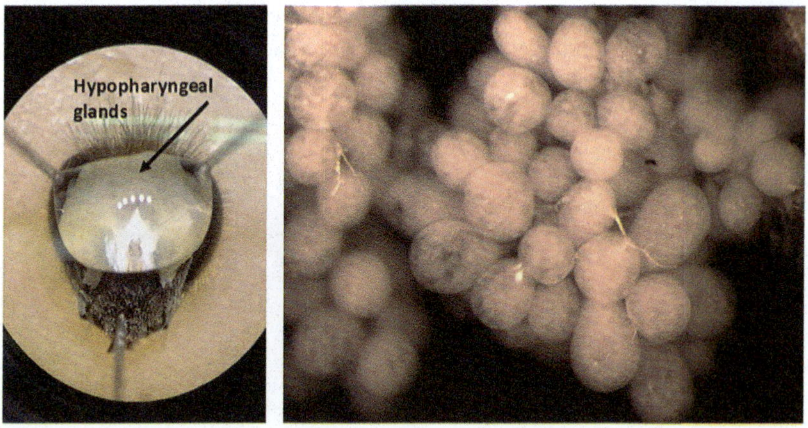

FIGURE 2.2 **The hypopharyngeal glands in the head of a worker bee and dissected out of the head to show their structure.** The hypopharnygeal glands in the head of a worker honey bee. The picture on the right shows the glands removed from the bee. The hypopharnygeal glands are composed of grape-like clusters of sacs containing brood food.
Photos by Emily Watkins DeJong.

HPGs secrete a highly nutritious milky white liquid called "royal jelly." Royal jelly contains essential amino acids, lipids, vitamins, the neural transmitter acetylcholine, and other nutrients needed for the development of larvae into adults (see ref. [5]). Nurse bees feed larvae royal jelly for the first 48 h after hatching from eggs. If larvae developing from fertilized eggs are fed royal jelly for their entire larval life, they will develop into queens. Larvae weaned from royal jelly after the first 48 h and fed a mixture of honey, bee bread (a mixture of pollen and nectar), and royal jelly, called "worker jelly," develop into workers [6–8]. Nurse bees also feed worker jelly to the queen and adult bees. After workers are about 7 days old, they consume less pollen and their HPGs no longer secrete royal jelly. At this point, the workers no longer care for brood and begin performing other tasks in the hive (see ref. [9]).

When a queen lays an egg, it takes three days for it to hatch into a larva (Fig. 2.3). Worker larvae take 6 days to mature and enter the pupal stage. Comb cells where bees are pupating are sealed until the fully developed adult releases itself by chewing through the seal on the top of the cell. Pupation and emergence from the capped cell occur over 12 days. Therefore, the development time for a worker bee from egg to adult is 21 days [10]. Development times are consistent and predictable because the bees actively maintain their nest environment at about 33°C with 80% humidity [11].

For the colony to grow, adult workers must live for more than 21 days. From spring until fall, workers spend 1–2 weeks after emergence performing duties in the hive. The young workers first serve as "nurse bees" that care for the larvae. After brood care, the workers shift to various in-hive duties such as caring for the queen, receiving and storing nectar loads from incoming foragers, building comb, removing debris and dead bees from the hive, and

FIGURE 2.3 **Honey bee eggs, larvae, and sealed brood in comb cells.** Immature stages of worker honey bees. The eggs are small, white, and oblong and are at the bottom of the cells. Larvae are white and curled into cells. The larvae grow and eventually fill the cell (mature larva). The capped cells contain pupae.
Photographs by Emily Watkins DeJong.

guarding the colony from intruders especially at the hive entrance. When workers become about 3 weeks old, they begin foraging. Depending upon the distance the bees have to fly to obtain nectar, pollen, and other resources such as water or propolis, foragers can live for a week or more.

The series of tasks required for colony survival and performed by bees of different ages is referred to as "the division of labor" [12]. The division of labor can help insure that workers live longer than 21 days because if the worker spends 3 weeks in the hive and a week or more foraging, their lifespan is at least 28 days, enabling the colony to grow.

The transition from in-hive tasks to foraging requires physiological and behavioral changes in the bee that are mediated by a compound called juvenile hormone (JH) [13]. Bees performing in-hive tasks have low JH levels compared to foragers where JH levels are high. Another compound that regulates division of labor, aging, and production of brood food is vitellogenin (Vg) [14–17]. Vg is synthesized in the fat bodies and released into the hemolymph at high levels in nurse and winter bees and at lower levels in foragers [15,18]. The HPGs of nurse bees absorb Vg from the hemolymph and metabolize it into amino acids that are used during the synthesis of royal jelly.

2.3 Drones

Colonies produce drones at the extremes of two possible events. Either the queen is running low on sperm and is laying unfertiled eggs or the colony is thriving and has enough extra resources to produce drones. Strong thriving colonies begin rearing drones in spring and early summer when nectar and pollen are abundant and colonies are building. Some drones also can be produced throughout the summer and early fall in large colonies. Rearing drones is energetically expensive for the colony, especially since they serve no known function other than mating with queens from other colonies. However, drones are an investment in the future of honey bee populations. Genetic traits that enable colonies to thrive are passed on to new colonies through drones mating with virgin queens. Without drones to mate with, colonies would perish when they lose their queen and eventually honey bees would vanish across landscapes.

Drones require 24 days to develop into adults and another week of more to become sexually mature (see ref. [19]). Sperm production in drones begins during the larval stage and is completed during the pupal stage. Drones do not produce sperm as adults. There can be 3.6–12 million sperm in drone semen. However, drones may have fewer viable sperm because of the health and genetic composition of the colony, diet, season, pesticide exposure, miticides used for Varroa control, pathogens, and parasites like Varroa (see refs [19,20]).

The lifespan of drones varies depending on the season. In all cases, drones remain in the brood area of the hive during the first days after emergence, and are fed and groomed by workers. After about a week, drones take orientation flights to learn landmarks and nest locations. As the drone matures, he will find drone congregation areas (DCAs) where thousands of drones from many colonies have come together to attract virgin queens. Drones fly short distances to find DCAs so they can spend more time in them and also avoid mating with virgin queens from their own colonies that fly several kilometers away from the hive to mate. Odors emitted by the drones probably aid in the formation of DCAs and affect interactions among the drones. When a virgin queen enters a DCA, drones detect her by visual cues and the pheromones she releases. Queens mate with 12–14 drones and in some cases many more in DCAs, ensuring genetic diversity in the colony [21]. Mating is a lethal act for drones, as they die soon after copulation.

2.4 Yearly cycle of honey bee colonies

Honey bee colonies have a yearly cycle that begins in mid-winter when egg laying resumes from a decline that began the previous fall [22]. In temperate areas, brood rearing can begin while the bees are still confined in the hive in their winter cluster. The number of eggs that are laid and reared to adults depends upon the health of the queen, numbers of adult worker bees, and food stores. Colonies can begin rearing brood before flowering plants are available by using honey and pollen collected during the previous fall. The resources could have been stored in comb cells and in the workers themselves, particularly in their fat bodies [23].

When enough worker bees survive the winter so that there are some to rear brood and others to forage for nectar and pollen, brood rearing can increase, and the colony can build throughout the spring and early summer. However, there are instances where weather might not be favorable for foraging or flowering plants are scarce so that the colony experiences nutritional stress. The stress activates a feedback loop that slows brood rearing by switching bees from in-hive tasks to foragers at younger ages [24]. Fewer brood are reared as more bees are recruited to forage for nectar and pollen. If more flowers bloom and weather is favorable for flight, the larger foraging population enables the colony to gather more resources and alleviate the nutritional stress. With more resources especially pollen brought into the hive, younger workers resume brood rearing, and the colony begins to grow again. However, if resources continue to be limited so less brood is reared, colonies can begin the downward spiral that ends in colony loss in early spring.

As spring turns to summer, and nectar and pollen continue to be available, the queen will lay increasing numbers of eggs and colonies will grow especially when resources are abundant. Maximizing colony growth in spring and early summer is essential for colony reproduction (i.e., swarming), drone production and successful overwintering. Large colonies also have greater capacity to adjust to changing environmental conditions because there are more workers to regulate temperature and humidity in the hive. A large colony also can collect enormous amounts of resources and store them in the hive to use when flowering plants are unavailable.

As summer ends, and fall approaches, egg laying and brood rearing decline, and in temperate regions, comes to an end in October or November. As brood rearing slows, pollen and nectar collected from fall flowers are stored in the colony or consumed by workers and stored in their fat body. The resources are used during winter when bees are confined in the hive. Confinement occurs when ambient temperatures drop below 50°F (10°C). Bees respond to low ambient temperatures by forming a tight cluster in the hive that can be thermoregulated to temperatures that enable the bees to remain active (90°F–95°F or 30°C–35°C). Changes in the physiology of individual bees also occur in the fall, enabling the workers to transition from summer into "winter bees" [23,25].

The development of winter bees is an example of how shifts in the physiological systems of individuals scale up to create colony-level effects. Winter bees differ from bees in spring and summer and are better equipped physiologically for extended periods of confinement in the hive. For example, winter bees might have larger fat bodies [26] with higher concentrations of lipids [27] than summer bees. Winter bees are heavier than summer bees and have higher amounts of total proteins, glucose, glycogen, and lipids [28,29] and lower JH titers [23,29–31]. Winter bees also have greater longevity than summer bees [32]. While summer bees usually live for about a month, winter bees can live for 5 months or more.

The cues that initiate the transformation of summer to winter bees are not well understood, but there are corollaries with seasonal conditions. For example, the transition to winter bees is associated with the reduction and eventual end of brood rearing in the fall that is correlated with shorter daylength and lower temperatures. However, neither factor alone can cause the transition from summer to winter bees (see ref. [25]). Determining if warmer fall temperatures associated with climate change affect the formation of winter bees and overwintering colony survival are areas requiring further study.

While the colony is still confined in the hive in late January or early February, the bees begin rearing brood. Brood rearing in mid-winter is essential for colony growth in the spring because the workers that emerge will replace the overwintered bees that die shortly after foraging resumes in the spring. The rate that overwintered bees die depends on weather conditions being favorable for foraging and the ages and longevity of bees in the overwintering population. If most bees in the overwintering population live long enough to be replaced by new adults from the first brood cycles, the colony will slowly begin to grow as it comes out of winter. However, if the winter cluster has a majority of short-lived bees, when foraging resumes those workers may die at rates that exceed the emergence of new adults to replace them. A decline in colony populations is common in the spring and is referred to as "spring dwindling." If there are enough brood and emerging adult bees, the colony can grow out of spring dwindling. However, colonies comprised of workers with short lifespans due to disease or poor nutrition experienced during the previous fall may not be able to rear enough brood to grow out of spring dwindling and may perish in the spring [33,34].

Healthy well-fed colonies will grow throughout the spring and summer by the simple system of queen egg laying, care of brood by young bees, foraging by older bees, and adult worker lifespan exceeding worker development time. However, stress factors can disrupt the system particularly if they shorten worker lifespans. Stress from poor nutrition or pathogens influences the physiological systems of individual bees, and the effects scale up to influence the entire colony. Colonies under stress do not grow nor rear drones or become large enough to swarm. Ecosystems where colonies reside also are affected. Fewer bees can limit pollination and seed set, further reducing flowering plants for future generations of bees. Colonies that do not rear drones also affect future generations. Queens need to mate with numerous drones to create a colony with the genetic diversity it needs to be healthy and grow. So, the effects of stress are both immediate and obvious and long term and pernicious. Landscapes that are supportive of pollinators can change the momentum into a positive direction. Well-fed colonies are more resilient to stress, and having sufficient resources to grow, they will produce drones and swarms that can increase the numbers of pollinators and flowering plants across landscapes.

2.5 Swarming—the reproductive system of the colony

In spring and early summer, colonies grow because of increased brood rearing especially if food is plentiful. Many plant species such as flowering fruit trees have massive blooms (sometimes called "big bang" blooms) in the spring with high numbers of flowers available in relatively small areas. The blooms usually occur for just a few weeks, but create strong nectar and pollen flows. The large amounts of resources brought into the hive by foragers coupled with increased brood rearing can cause a lack of comb space. In managed colonies, beekeepers will add more wax combs to let the colony expand. In feral colonies that are housed in cavities of limited space, the overcrowding will initiate swarming.

Swarming most often occurs during May and June in northern temperate regions, although it also can occur later in the summer. Swarms that occur in the spring have the best chance of surviving through the winter, as there is sufficient time for the colony to grow and gather resources. The swarming process begins with workers rearing new queens. There also will be fewer workers feeding the queen each day, and her egg laying will decline. The reduction in food to the queen occurs several days after the bees begin rearing new queens. Colonies preparing to swarm will rear many queens in individual cells called 'swarm cells'. The swarm cells are lined up across the bottom of combs containing brood. When the first queen is ready to emerge from her cell, the workers become agitated and fly around the entrance of the hive. Other bees consume honey in preparation for the swarm's departure. Signals that workers use to organize and decide when to leave the hive are not well understood, but what has been observed is the activity around the colony's queen. Just prior to leaving the hive, the queen is surrounded by workers that push her out of the hive. Several thousand workers join the queen outside the hive to form the swarm. The swarm flies away from the hive and usually lands somewhere close to where it originated. After landing, workers in the swarm detect the queen, and cluster around her (see ref. [35]).

The cluster of bees needs to find a new permanent nesting site. A small group of workers of foraging age (scouts) leave the swarm cluster and search for a new nesting cavity [36]. The scouts return to the cluster and communicate the location of potential nesting cavities using a dance language. From the many dances that might be occurring on the cluster, the bees reach a consensus on a single location for the new nest site. At that point, the swarm takes off and follows several guide bees to the new colony location [37].

When bees find a new nesting site, the colony is teetering on extinction from two major challenges: (1) allocation of resources for comb building, foraging, and brood rearing and (2) maintaining age structure for colony growth. When bees leave their natal nest, the only resources they have are what they are carrying in bodies. The resources, primarily glucose, need to be divided into those used for comb building so there is space for the queen to lay eggs, food for the bees, and fuel for foragers. Wax comb is a nonliquid asset. Once the sugars are converted to wax, they cannot be converted back to sugar. Too great an investment in wax will starve the colony. Too little investment could limit space for storing resources and for brood rearing, thus slowing colony growth and reducing the chances of survival, especially over the winter (for a review of colony-level control of comb building (see ref. [38])).

Maintaining an age structure so the population can grow is another challenge for a newly established colony. The success of a swarm is measured by the ability to grow a population of sufficient size and collect enough resources to overwinter and produce a new swarm the following year. When the swarm establishes a new nest, they immediately begin building comb. The queen will start laying eggs in 2–3 days. Remember, it takes 21 days for a worker egg to develop into an adult. Therefore, the swarm population will decline for the next 23–25 days or more before new workers emerge to replace the original workers in the swarm. It is imperative then that swarms find nesting sites near large expanses of flowering plants to extend the life of the foragers and provide resources to feed the colony, rear brood, and build comb.

2.6 Genetic diversity—modulating colony system responses

Well-designed systems are capable of modulating their responses when conditions change. Responses are graded rather than oscillating between highs and lows [39]. The key to modulated responses in honey bee colonies is genetic diversity among the worker population. Colonies have populations that are genetically diverse because queens mate with numerous drones. Queens usually take a single mating flight as the survival of the colony hinges on her thorough mating and safe return. By thorough mating, I mean copulating with many drones. A well-mated queen will have semen from a dozen or more drones and will use the semen to fertilize eggs and produce workers for the remainder of her life.

Queens mate with many drones, so a colony is composed of numerous patrilines (genetic information obtained from the father) and a high degree of genetic diversity. The diversity can be expressed in various ways but particularly in how workers respond to environmental conditions. Based on their patriline, some workers might have a greater probability of performing a certain task in response to conditions compared with workers of other patrilines [40]. Bees with a high likelihood of performing one task might also be more likely to perform a similar task and thus have a repertoire of behaviors based on their patriline [41]. For example, workers that are specialists in removing diseased brood (hygienic specialists) also have a geater tendency to remove dead bees from the hive (undertaker specialists).

Whether a worker has a greater tendency to perform a specialized task depends on their threshold for detecting a stimulus for the task (e.g., an odor emanating from a diseased larva), and then responding to it with a behavior (removing the diseased larva) [42]. The degree of stimulation workers need (i.e., the threshold) before performing a task varies with patriline. Individuals with a particular patrilines might require a less stimulus before responding to some tasks (i.e. low threshold) and a others require higher stimulus (high threshold). Differences among patrilines in the thresholds required to respond to a stimulus and then perform a behavior provide a colony with temperance and flexibility in how it responds to changes in the environment. For example, a colony might have a low-level brood infection and odors emanating from the diseased larvae. The odor is detected by workers of a patriline with a low response threshold, and they remove the larvae. Workers with other patrilines having higher thresholds are free to perform other tasks as required by the colony or in response to stimuli where they have a low threshold. This could be guarding behavior or nest defense. However, if the brood disease spreads to many larvae so that the odor is stronger, bees with higher response thresholds will be activated and join in to remove the diseased larvae. Thus, depending on the intensity of the stimulus, the colony can modulate its response so that the behaviors performed by the workforce are spread across numerous tasks but, if need arises, can be concentrated on those requiring the most immediate attention.

2.7 Division of labor— the system can adjust

For a honey bee colony to grow and survive, numerous subsystems need to be coordinated so that workers can respond to conditions with appropriate behaviors and levels of participation. Resources must be gathered and stored, young bees need to be reared, comb must be built, and the nest needs to be kept clean and defended from pillaging invaders. Worker bees perform tasks at different ages, but based on factors such as interactions with nestmates, genetic inclinations, or the amount of food in the colony, workers can shift to tasks requiring more immediate attention. The flexibility of individual worker behaviors scales to colony-level adjustments to the current environmental conditions.

The behavioral flexibility within each worker and its expression due to interactions with nestmates was demonstrated in a classic study involving young nurse bees and older workers that were foragers. Workers that had just emerged from cells after pupation were removed from their colony and isolated as a group of same-aged bees. A second set of newly emerged workers was marked so they could be identified and were released back into a colony. In the first group, workers interacted only with others of the same age, while those released in the hive interacted with bees of all ages. JH levels were measured in the bees as an indicator of their aging process and potential behaviors they would perform. Remember, young bees that perform nurse behaviors have low JH, while older bees that forage have higher JH levels. The young bees that were put back into their colony had low JH levels. However, some workers in the group that were isolated and interacting only with workers of the same age developed JH levels comparable to 24-day-old foragers. The changes into forager JH levels occurred within a week. The bees began foraging when they were only 13–16 days old and were able to orient to their hive and return with pollen loads. Workers that become foragers earlier in their adult lives are called "precocious foragers." A second part of the study involved adding older bees that were foraging or defending the nest to the group of young bees. Interacting with the older bees inhibited JH production, and precocious foraging in the younger bees did not occur [43].

How quickly workers transition from nurse bees to foragers also can be affected by how available food is in the colony. Nurse bees differ from foragers in that nurses consume pollen and have higher lipid stores in their abdominal fat body tissue than foragers [44]. Interactions mediated by chemical cues (pheromones) exchanged between foragers and nurse bees and between brood and nurse bees affect nervous system processes and cause an increase in nutrient storage in the fat body and HPGs in nurse bees. Foragers, in contrast, have decreased expression of genes related to lipid and protein storage compared with nurse bees and increased expression of genes related to carbohydrate metabolism needed for flight. Nutritional stress that limits pollen consumption reduces fat stores in younger workers. Lower fat stores accelerates worker transition from house bee to forager in an attempt to boost incoming food and relieve nutritional stress (see ref. [45]).

A colony can rebuild its foraging population and avert starvation through the accelerated maturation of young bees into precocious foragers, but there is a price to pay in colony growth. If bees become foragers when they are younger than 14 days old and forage for a week, their lifespan is 21 days. Since it takes 21 days for an adult worker to develop, the colony will not grow. Starting to forage at a younger age or for a shorter time will cause the colony to decline. Precocious foraging is a short-term adjustment to advert starvation, but if prolonged will ultimately reduce brood rearing and cause the colony to decline and perish. The scenario of precocious foraging and its effects on colony survival will come up again in the context of diseases, parasites, nutritional stress, and climate change.

2.8 Immune systems

A healthy colony has thousands of bees living in close quarters in a hive where the climate is kept warm and moist. The bees are in constant contact, and are exchanging food mouth-to-mouth (trophallaxis). The conditions are conducive for colony cohesiveness, but also ideal for the growth and spread of pathogens among the bees especially in larger colonies. Pathogens can enter colonies in several ways. If foraging areas are shared with bees carrying pathogens from other colonies, flowers can be hotspots for disease transmission (see ref. [46]). The microscopic parasite, *Nosema ceranae* is spread this way [47]. Workers and drones also can drift into colonies that are not their own, and transmit the pathogens they might be carrying.

Most honey bee pathogens, such as bacteria, fungi or viruses, share a long evolutionary history with bees. The pathogens probably started out having commensal (i.e., one side, either the bee or the microorganismbenefited, while the other side was unaffected) or even symbiotic (both sides benefitted) relationships with bees. At some point however, changes in key biotic or abiotic factors could have greatly increased the replication of the microorganism or caused compromised metabolic functions in the bee. The changes could result in pathologies, such as deformed wings, brood mortality, or reduced lifespans in adults. Clear symptoms of an infection that result from an imbalance in host–microbial interactions are what we call "diseases."

Pathogens and their hosts are in a continuous arms race where adjustments on one side prompts the other to adapt so it remains successful. Honey bee colonies take a multisystem strategy to limit the growth of parasites and pathogens. Certain components emerge from hive structure and colony organization, but there also are group behaviors and the immune system in individual bees. Working together, the different lines of defense against disease generate a social rather than an individual immunity [48,49].

If we begin with the outer most line of defense against pathogens, it would be coating the hive cavity with the antimicrobial resins from plants. Bees forage for plant resins and bring them back to the hive where the resins are mixed with bee saliva and beeswax to form "propolis." Propolis contains terpenoids, steroids, sugars, amino acids, and bioactive compounds such as flavonoids and phenolic acids and their derivatives [50]. The exact composition of propolis differs depending on the plant source, geographical location, and collection season. In general, however, propolis contains compounds that are antibacterial, antifungal, antiprotozoan, and possibly antiviral (most studies have been done with human rather than honey bee viruses). Bees spread propolis throughout the hive to create a waterproof antimicrobial envelope surrounding the colony. Bees also use propolis to entomb animals or insects that have died in the hive and are too large for the bees to remove. The collection of plant resins to coat the hive originated from honey bees' natural nesting sites that often were cavities in trees. The bees would remove dead wood from the cavity and then coat the cavity with propolis to fill cracks, prevent the wood from rotting (important because the environment in the hive is warm and moist) and keep the hive dry.

In addition to the structure of the hive, the colony organization provides social immunity through the division of labor. Bees of similar age are found together in common locations in the nest performing similar tasks. Young bees caring for brood are at the center of the nest and older workers that are storing honey, guarding the colony, or foraging are on the outermost combs. Workers on brood frames caring for larvae have minimal contact with older workers that are guarding the nest entrance or foraging and might bring pathogens into the colony. The oldest workers that usually are foragers often have the highest titers of pathogens such as viruses and Nosema. Most foragers die outside the hive and in the process remove pathogens from the colony [48].

Going into the colony population, the next component in colony-level immunity is removal of dead and diseased bees and larvae. Workers that have aged out of nursing activities, but are not old enough to be foragers, usually are the bees that remove dead and diseased nestmates. Removing diseased brood is called "hygienic behavior" and removing dead adults is called "undertaking or necrophoric behavior." There is a genetic predilection for hygienic behavior as described above, and bees can be bred for high sensitivity to odors emanating from diseased brood. Colonies with high numbers of hygienic bees will detect and remove infected larvae when pathogens are still at low levels, enabling the colony to be resistant to brood diseases like American foulbrood and chalkbrood (reviewed in refs [51,52]).

Colony structure and group behaviors can reduce pathogen levels, but if these systems fail, individual bees have their own immune systems that can react to pathogens. Individual immunity begins at the bee's exoskeleton that serves as a water tight barrier against parasites and pathogens. In insects and other arthropods, the outermost layer of the exoskeleton (epicuticle) is composed of wax and other waterproof compounds. Beneath the epicuticle are microfibers of chitin surrounded by globular proteins that create a strong yet flexible outer surface for the bee. When insects molt, they shed their cuticle and form a new one beneath. The new cuticle is light colored until it hardens (sclerotization) and becomes pigmented through a process of melanization.

Melanization also can occur if the insect is wounded or is infected with a virus. When injury or infection occurs, the bee responds by mobilizing cells called hemocytes that carry melanin to the injured or infected site. The hemocytes lyse and release the melanin to patch the wound and prevent infection [48].

Pathogens can bypass physical and chemical barriers by entering their hosts through an oral route or by partnering with parasites like Varroa mites that can transmit pathogens during feeding. If a pathogen enters the body, honey bees can protect themselves from infection with cellular and humoral (i.e., antibody related) immune responses [53]. The responses are activated by recognition of molecules on the surface of pathogens. The specific structures are called pathogen-associated molecular patterns (PAMPs), and are recognized by pattern recognition receptors (PRRs) that are proteins that can detect PAMPs [54]. When a PAMP is recognized by the PRRs, it is chemically bound and triggers signaling cascades that lead to the activation of hemocyte-mediated cellular immune responses. The responses include phagocytosis, a cellular process for ingesting and eliminating microorganisms and other particles larger than 0.5 μm [55]. Hemocytes also can initiate nodule formation and encapsulation of the invading pathogens. The melanin contained in the hemocytes is used for the encapsulation. The hemocytes also can begin the synthesis of antimicrobial peptides (AMPs). Peptides are short chains of chemically connected amino acids. Several AMPs such as abaecin, apidaecin, hymenoptaecin, and defensin have been identified in the hemolymph of honey bees with microbial infections [56]. AMPs can prevent infection from a wide range of microorganisms (bacteria, fungi, some parasites, and viruses) by compromising the structure of a pathogen's membrane or inhibiting protein, RNA, or DNA synthesis.

Although AMPs can be effective against viruses, recent studies indicate that RNA interference (RNAi) is the major antiviral innate immune response in honey bees and other insects. RNAi is an innate antiviral pathway that is triggered by the detection of double-stranded RNA (dsRNA) from the virus generated during RNA virus replication. The RNAi response includes the release of an RNase III-like enzyme called Dicer 2 (Dcr2) that recognizes virus dsRNA as a PAMP and cleaves long stretches of it into short interfering RNAs (siRNAs) essentially halting virus replication (for more information see ref. [54]).

Immunity to pathogens can be passed down from the queen to her offspring through a process called "transgenerational immune priming." Transgenerational immune priming is commonly found in insects and other invertebrates and enables an immune response to occur quickly when there is exposure to a pathogen [57]. The priming begins when females transfer ingested fragments of pathogens to their developing eggs. The fragments (immune elicitors) act like a vaccine and can elicit an immune response in the embryo that enables a more efficient response if the pathogen is encountered again. In honey bees, Vg may facilitate transgenerational immune priming. Vg is present in eggs, HPGs of nurse bees, and brood food. Vg appears to transport immune elicitors, created when the queen transfers ingested pathogen fragments and PAMPS to their developing eggs.

Immune priming by the queen to her offspring requires that the queen encounters the same pathogens as workers and drones. How could this happen, since the queen has limited exposure to environmental pathogens? She leaves the nest for a mating flight, and with the exception of swarming, the queen may never leave again. So, what is the source of pathogen fragments and PAMPS that are transferred to eggs? In a study where heat-killed bacteria (Escherichia *coli*) were fed to worker bees, bacterial fragments were transferred to the HPGs of the workers. Vg played a role in the transport process [58]. Whether the bacterial fragments bound to the HPGs are incorporated into royal jelly fed to the queen or the fragments induce changes to jelly composition is not known. Royal jelly contains several components with immunological functions, including defensin 1 and major royal jelly protein 3. The presence of bacterial fragments in the HPGs might stimulate the production of more immunomodulators to be incorporated into the jelly, but this requires further study. However, the immune priming is initiated, the system enables future generations to be prepared for pathogens encountered by their predecessors.

2.9 Thermoregulation— not too hot, not too cold

Honey bees, like other insects, are ectotherms (also called poikilotherms or cold-blooded animals). Ectotherms use heat from the environment to initiate their metabolic processes. When temperatures are low, ectotherms move slowly or not at all. If temperatures get too high, ectotherms can die from overheating. Honey bee colonies have adapted systems to regulate hive temperatures during cold and hot weather [59]. Temperature regulation (often called thermoregulation) is important not only for survival but also for rearing brood and optimizing adult activity and metabolism. Because honey bees can thermoregulate the hive, they are one of the few temperate zone insects that are active throughout the year.

Temperatures in colonies need to be maintained between 32°C–36°C (89.6°F–96.8°F) and optimally at 35°C (95°F) for brood to develop normally (see Ref. [60]). At lower or higher temperatures, brood will perish or emerge from brood cells with abnormalities such as shriveled wings, and smaller adult body size possibly due to accelerated development. Workers also show cognitive deficiencies possibly from neurological damage and can begin foraging earlier than expected in adult life [61,62]. The pupal stage seems to be especially sensitive to temperature extremes.

Strategies to heat and cool the hive revolve around the distribution of bees on the combs. When outside temperatures are high, workers spread out across the combs, and the entrance and front of the hive and fan their wings sometimes over droplets of water collected by foragers. The fanning creates air circulation and accelerates water evaporation much the same as an evaporative cooler and fans do in our homes. The detection of high temperatures in the hive and the initiation of fanning appear to be based on patriline much as the hygienic behavior described above. Experimental colonies established with a multiply-mated queen (the colony had many patrilines) and exposed to high temperatures had a gradually increasing number of bees respond by fanning to cool the hive. Fanning began with workers having low thresholds and was followed by workers with higher thresholds generating a graded response to the high temperatures. In contrast, colonies headed by queens mated to a single drone (single patriline colonies) responded to high temperatures with high numbers of fanning bees. The fanning caused temperatures to oscillate between being too low and too high. There was less variation in temperatures in colonies with multiple patrilines compared with those having single patrilines [60].

To remain active throughout the year, honey bees need to regulate temperatures in the hive during the cold of fall and winter. When ambient temperatures drop below 15°C, workers begin to cluster together around the queen on the combs comprising the brood nest. The bees generate heat by shivering the flight muscles in their thorax [63]. Flight muscles in insects are among the most metabolically active tissues known to occur in animals, and can efficiently convert chemical energy (i.e., food) to mechanical power. The by-product of the conversion is heat. The densely distributed plumose body hairs on the thorax of bees may also trap the heat generated from the flight muscle.

Older bees that have better developed flight muscles than younger workers are on the outer layers of the winter cluster, while younger bees are at the center of the cluster [64]. The tightness of the cluster is greatest on the outside layers. The cluster loosens toward the center so that bees and the queen can move easily and air can be exchanged. Since contacts among the bees is greatest where the cluster is tightest, thermal conductivity is greatest on the outside areas of the cluster.

As daytime temperatures warm above 15°C, the cluster of bees will become looser, and some bees will fly out of the hive to defecate. These are called "cleansing flights." Warming temperatures that enable bees to fly for perhaps a few hours a day in the late winter or early spring can reduce pathogens such as Nosema that reproduces in the honey bee gut. However, if flowers are not available, bees will be fueling the flights with resources they may need to keep the colony warm if the weather turns cold.

2.10 Conclusions

In this chapter, I provided a broad overview of honey bee biology and how interactions among the bees that participate in different subsystems of the colony enable the population to grow and adjust to changing environmental conditions. The subsystems arise from basic elements of individual bees such as their genotype, age, or physiology and shape the interactions among individuals of similar disposition that culminate in the performance of a specific task (e.g., comb building) by groups of bees. At any time, there may be multiple subsystems in operation that are interconnected to create a system that can respond and adapt to environmental conditions throughout the year. In this chapter, I focused on particular subsystems that are vulnerable to stresses from climate change. Systems work within a particular range of conditions where they can adjust and continue to function. Climate change can cause environmental conditions that are outside the ranges where colonies can adjust and lead to colony loss from the cascading effects of subsystem failures. The following chapters will describe how factors associated with climate change are causing colony losses due to breakdowns in the subsystems described above. I also will provide some solutions at landscape and management levels that could reduce stress on subsystems and enable them to work within the ranges that ensure colony growth and survival.

References

[1] Naumann K, Winston ML, Slessor KN, Prestwich GD, Webster FX. Production and transmission of honey bee queen (Apis mellifera L.) mandibular gland pheromone. Behav Ecol Sociobiol 1991;29(5):321–32.

[2] Hoover SE, Keeling CI, Winston ML, Slessor KN. The effect of queen pheromones on worker honey bee ovary development. Naturwissenschaften 2003;90(10):477–80.

[3] Maisonnasse A, Lenoir J-C, Beslay D, Crauser D, Le Conte Y. E-b-Ocimene, a volatile brood pheromone involved in social regulation in the honey bee colony (Apis mellifera). PLoS ONE 2010;5(10):e13531.

[4] Gilley DC, DeGrandi-Hoffman G, Hooper JE. Volatile compounds emitted by live European honey bee (Apis mellifera L.) queens. J Insect Physiol 2006;52(5):520–7.

[5] Drapeau MD, Albert S, Kucharski R, Prusko C, Maleszka R. Evolution of the Yellow/Major Royal Jelly Protein family and the emergence of social behavior in honey bees. Genome Res 2006;16(11):1385–94.

[6] Wang Y, Ma L, Zhang W, Cui X, Wang H, Xu B. Comparison of the nutrient composition of royal jelly and worker jelly of honey bees (Apis mellifera). Apidologie 2016;47(1):48–56.

[7] Crailsheim K. The protein balance of the honey bee worker. Apidologie 1990;21(5):417–29.

[8] Haydak MH. Honey bee nutrition. Annu Rev Entomol 1970;15:143–56.

[9] Ahmad S, Khan SA, Khan KA, Li J. Novel insight into the development and function of hypopharyngeal glands in honey bees. Front Physiol 2021;11:615830.

[10] Page REJ, Peng CY. Aging and development in social insects with emphasis on the honey bee, Apis mellifera L. Exp Gerontol 2001;36:695–711.

[11] Abou-Shaara HF, Owayss AA, Ibrahim YY, Basuny NK. A review of impacts of temperature and relative humidity on various activities of honey bees. Insectes Sociaux 2017;64(4):455–63.

[12] Johnson BR. Division of labor in honeybees: form, function, and proximate mechanisms. Behav Ecol Sociobiol 2010;64:305–16.

[13] Huang ZY, Robinson GE, Tobe SS, Yagi KJ, Strambi C, Strambi A, et al. Hormonal regulation of behavioural development in the honey bee is based on changes in the rate of juvenile hormone biosynthesis. J Insect Physiol 1991;37(10):733–41.

[14] Amdam GV, Norberg K, Hagen A, Omholt SW. Social exploitation of vitellogenin. Proc Natl Acad Sci U S A 2003;100(4):1799–802.

[15] Amdam GV, et al. Altered physiology in worker honey bees (Hymenoptera: Apidae) infested with the mite Varroa destructor (Acari: Varroidae): a factor in colony loss during overwintering? J Econ Entomol 2004;97(3):741–7.

[16] Seehuus SC, Norberg K, Gimsa U, Krekling T, Amdam GV. Reproductive protein protects functionally sterile honey bee workers from oxidative stress. Proc Natl Acad Sci U S A 2006;103(4):962–7.

[17] Münch D, Ihle KE, Salmela H, Amdam GV. Vitellogenin in the honey bee brain: atypical localization of a reproductive protein that promotes longevity. Exp Gerontol 2015;71:103–8.

[18] Corona M, Velarde RA, Remolina S, Moran-Lauter A, Wang Y, Hughes KA, et al. Vitellogenin, juvenile hormone, insulin signaling, and queen honey bee longevity. Proc Natl Acad Sci U S A 2007;104(17):7128–33.

[19] Rangel J, Fisher A. Factors affecting the reproductive health of honey bee (Apis mellifera) drones—a review. Apidologie 2019;50(6):759–78.

[20] Yániz JL, Silvestre MA, Santolaria P. Sperm quality assessment in honey bee drones. Biology 2020;9(7):174.

[21] Withrow JM, Tarpy DR. Cryptic "royal" subfamilies in honey bee (Apis mellifera) colonies. PLoS ONE 2018;13:e0199124.

[22] DeGrandi-Hoffman G, Gage SL, Corby-Harris V, Carroll M, Chambers M, Graham H, et al. Connecting the nutrient composition of seasonal pollens with changing nutritional needs of honey bee (Apis mellifera L.) colonies. J Insect Physiol 2018;109(June):114–24.

[23] Döke MA, Frazier M, Grozinger CM. Overwintering honeybees: biology and management. Curr OpInsect Sci 2015;10:185–93.

[24] Brodschneider R, Crailsheim K. Nutrition and health in honey bees. Apidologie 2010;41(3):278–94.

[25] Mattila HR, Harris JL, Otis GW. Timing of production of winter bees in honey bee (Apis mellifera) colonies. Insectes Sociaux 2001;48(2):88–93.

[26] Koubová J, Sábová M, Brejcha M, Kodrík D, Čapková Frydrychová R. Seasonality in telomerase activity in relation to cell size, DNA replication, and nutrients in the fat body of Apis mellifera. Sci Rep 2021;11(1):1–12.

[27] DeGrandi-Hoffman G, Corby-Harris V, Graham H, Watkins deJong E, Chambers M, Snyder L. The survival and growth of honey bee (Hymenoptera: Apidae) colonies overwintered in cold storage: the effects of time and colony location. J Econ Entomol 2023;116(4):1078–90.

[28] Kunert K, Crailsheim K. Seasonal changes in carbohydrate, lipid and protein content in emerging worker honeybees and their mortality. J Apicul Res 1988;27(1):13–21.

[29] Kunc M, Dobeš P, Hurychová J, Vojtek L, Beani Poiani S, Danihlik J, et al. The year of the honey bee (Apis mellifera L.) with respect to its physiology and immunity: a search for biochemical markers of longevity. Insects 2019;10(8).

[30] Huang Z-Y, Robinson GE. Seasonal changes in juvenile hormone titers and rates of biosynthesis in honey bees. J Comp Physiol B 1995;165:18–28.

[31] Mattila HR, Otis GW. Dwindling pollen resources trigger the transition to broodless populations of long-lived honeybees each autumn. Ecol Entomol 2007;32(5):496–505.

[32] Fakuda H, Sekiguchi K. Seasonal Change of the Honeybee Worker Longevity in Sapporo, North Japan, With Notes on Some Factors Affecting the Life-Span. Japanese J Ecol 1966;16(5):206–12.
[33] Betti MI, Wahl LM, Zamir M. Age structure is critical to the population dynamics and survival of honeybee colonies. R Soc Open Sci 2016;3(11).
[34] Rajagopalan K, DeGrandi-Hoffman G, Pruett M, Jones VP, Corby-Harris V, Pireaud J, et al. Warmer autumns and winters could reduce honey bee overwintering survival with potential risks for pollination services. Sci Rep 2024;14:5410.
[35] Rangel J, Seeley TD. Colony fissioning in honey bees: size and significance of the swarm fraction. Insects Soc 2012;59:453–62.
[36] Seeley TD, Morse RA, Visscher PK. The natural history of the flight of honey bee swarms. Psyche: A J Entomol 1979;86(2-3):103–13.
[37] Makinson JC, Beekman M. Moving without a purpose: an experimental study of swarm guidance in the Western honey bee, Apis mellifera. J Exp Biol 2014;217:2020–7.
[38] Pratt SC. Collective control of the timing and type of comb construction by honey bees (Apis mellifera). Apidologie 2004;35:193–205.
[39] Meadows DH. Wright D, editor. Thinking in systems. London, UK: Earthscan; 2008.
[40] Robinson GE. Regulation of honey bee age polyethism by juvenile hormone. Behav Ecol Sociobiol 1987;20:329–38.
[41] Perez AA, Johnson BR. Task repertoires of hygienic workers reveal a link between specialized necrophoric behaviors in honey bees. Behav Ecol Sociobiol 2019;73:123.
[42] Masterman R, Ross R, Mesce K, Spivak M. Olfactory and behavioral response thresholds to odors of diseased brood differ between hygienic and non-hygienic honey bees (Apis mellifera L.). J Comp Physiol A 2001;187:441–52.
[43] Huang Z-Y, Robinson GE. Honeybee colony integration: worker-worker interactions mediate hormonally regulated plasticity in division of labor. Proc Natl Acad Sci U S A 1992;89(24):11726–9.
[44] Toth AL, Kantarovich S, Meisel AF, Robinson GE. Nutritional status influences socially regulated foraging ontogeny in honey bees. J Exp Biol 2005;208(24):4641–9.
[45] Walton A, Toth AL. Nutritional inequalities structure worker division of labor in social insects. Curr Opin Insect Sci 2023;58:101059.
[46] Cremer S, Armitage SAO, Schmid-Hempel P. Social immunity. Curr Biol 2007;17:R693–702.
[47] Graystock P, Goulson D, Hughes WOH. Parasites in bloom: flowers aid dispersal and transmission of pollinator parasites within and between bee species. Proc R Soc B: Biol Sci 2015;282(1813):20151371.
[48] Evans JD, Spivak M. Socialized medicine: individual and communal disease barriers in honey bees. J Invert Path 2010;103:S62–72.
[49] Cremer S, Sixt M. Analogies in the evolution of individual and social immunity. Philos Trans R Soc Lond B Biol Sci 2009;364:129–42.
[50] Contieri LS, de Souza Mesquita LM, Sanches VL, Chaves J, Pizani RS, da Silva LC, et al. Recent progress on the recovery of bioactive compounds obtained from propolis as a natural resource: processes, and applications. Sep Purif Technol 2022;298:121640.
[51] Spivak M, Gilliam M. Hygienic behaviour of honey bees and its application for control of brood diseases and Varroa: Part II. Studies on hygienic behaviour since the Rothenbuhler era. Bee World 1998;79(4):169–86.
[52] Spivak M, Reuter GS. Resistance to American foulbrood disease by honey bee colonies Apis mellifera bred for hygienic behavior. Apidologie 2001;32(6):555–65.

[53] DeGrandi-Hoffman G, Chen Y. Nutrition, immunity and viral infections in honey bees. Curr Opin Insect Sci 2015;10:170–6.

[54] Brutscher LM, Flenniken ML. RNAi and antiviral defense in the honey bee. J Immunol Res 2015;2015(1):941897.

[55] Uribe-Querol E, Rosales C. Phagocytosis: our current understanding of a universal biological process. Front Immunol 2020;11:1066.

[56] Durand T, Bonjour-Dalmon A, Dubois E. Viral co-infections and antiviral immunity in honey bees. Viruses 2023;15:1217.

[57] Freitak D, Schmidtberg H, Dickel F, Lochnit G, Vogel H, Vilcinskas A. The maternal transfer of bacteria can mediate trans-generational immune priming in insects. Virulence 2014;5(4):547–54.

[58] Harwood G, Amdam G, Freitak D. The role of Vitellogenin in the transfer of immune elicitors from gut to hypopharyngeal glands in honey bees (Apis mellifera). J Insect Physiol 2019;112:90–100.

[59] Stabentheiner A, Kovac H, Mandl M, Käfer H. Coping with the cold and fighting the heat: thermal homeostasis of a superorganism, the honeybee colony. J Comp Physiol A 2021;207:337–51.

[60] Jones JC, Myerscough MR, Graham S, Oldroyd BP. Honey bee nest thermoregulation: diversity promotes stability. Science 2004;305(5682):402–4.

[61] Poot-Báez V, Medina-Hernández R, Medina-Peralta S, Quezada-Euán JJG. Intranidal temperature and body size of Africanized honey bees under heatwaves (Hymenoptera: Apidae). Apidologie 2020;51:382–90.

[62] Simone-Finstrom M, Foo B, Tarpy DR, Starks PT. Impact of food availability, pathogen exposure, and genetic diversity on thermoregulation in honey bees (Apis mellifera). J Insect Behav 2014;27:527–39.

[63] Stabentheiner A, Pressl H, Papst T, Hrassnigg N, Crailsheim K. Endothermic heat production in honeybee winter clusters. J Exp Biol 2003;206(2):253–8.

[64] Southwick EE, Heldmaier G. Temperature control in honey bee colonies. BioScience 1987;37(6):395–9.

Chapter 3

Systems that produce pollination and fruit set

3.1 Introduction

Before I tell you about the systems and relationships that produce some of the most beautiful and delicious foods we eat, I want to give you a high-level overview of plants so you can be amazed by what they do. I also hope to convey how all lifeforms that require oxygen for respiration rely on plants. We also rely on plants for food, fiber, and the materials we use for shelter. Pollination by bees plays a critical role in the seed set and propagation of plants in agricultural and unmanaged ecosystems. From the perspective of human nutrition, plants that rely on bee pollination comprise 87 of the 115 most important crops in the human diet [1].

Plants are a fundamental element of life on Earth because they can capture sunlight and convert the solar energy into sugars. Plants accomplish the sunlight to sugar conversion by combining carbon dioxide and water using a process called photosynthesis. Carbon dioxide from the air enters plants through microscopic pores called stomata located on leaves, stems, and flowers. Water is absorbed through the roots as are minerals such as nitrogen, phosphorus, potassium, and iron. The minerals transport nutrients throughout the plant and are used in the synthesis of compounds such as chlorophyll. Light energy from the sun is absorbed by chlorophyll. The green color of leaves, stems, and unripe fruit is from the chlorophyll contained in chloroplasts; a type of organelle housed in plant cells. Chloroplasts are the site of photosynthesis. Light energy is absorbed and stored by chlorophyll and with the addition of carbon dioxide and water are used to make glucose. Oxygen is a by-product of photosynthesis and is released through the stomata.

The sugars produced through photosynthesis are building blocks for more complex molecules such as starches, fats, oils, and cellulose that are combined to make wood, flowers, fruit, and seeds. When plants bind nitrogen to glucose, they make amino acids that are the building blocks of plant proteins. Plants can grow to extraordinary sizes (Fig. 3.1), produce beautiful and fragrant flowers,

The Honey Bee and Climate Change. DOI: https://doi.org/10.1016/B978-0-443-28995-8.00016-7

FIGURE 3.1 **Giant Sequoia (Sequoiadendron giganteum) at Calaveras Big Trees State Park, Arnold, California, USA.** A towering giant sequoia (*Sequoiadendron giganteum*) in Calaveris Big Trees State Park, Arnold, California, USA (https://www.parks.ca.gov/?page_id=551). Trees can be 250–300 ft tall (76–91 m). To provide perspective on the size of the trunk, the author is standing at the base. The tree is the product of photosynthesis where plants combine CO_2, sunlight, and water to create glucose.
Photographs by Gloria DeGrandi-Hoffman and Amelia D. Hoffman-Peiffer.

and set seeds encased in fruits using only sunlight, water, carbon dioxide, and minerals. That is amazing!

The process plants use to produce seeds and fruit begins with flowering. If you look at a field of wildflowers, you see great diversity in the shape, color, and structure of the flowers. Some flowers are large and showy, while others are small and subtle. Flowers are arranged on stems differently, too. Some plants like poppies (*Papaver* spp.) and lilies (*Lilium* spp.) have single flowers at the tip of a stem, while others like clovers (*Trifolium* spp.), dandelion (*Taraxacum* spp.), and sunflowers (*Helianthus* spp.) have groups of small flowers (florets) clustered together as inflorescences (Fig. 3.2). The florets can be so tightly packed together like on a sunflower (*Helianthus annuus* L.), that they appear to be a single large flower. The arrangement of flowers is important for attracting pollinators and moving pollen among the flowers.

Although the shape and colors of flowers and the way they are arranged on a stem might differ among plants, all flowers follow the same basic game plan when it comes to producing seeds. That is, pollen needs to be moved from the male to the female floral structures to initiate seed formation. Although many animals can transfer pollen among flowers, in this chapter, we will focus on the transfer of pollen by honey bees.

FIGURE 3.2 Plants the present flowers singly and in groups. Plants can present their flowers to pollinators singly as in the desert willow (*Chilopsislinearis* spp.) on the left or in clusters of florets as in the Rocky Mountain bee plant (*Cleome serrulata*) on the right.
Photographs by Emily Watkins deJong.

3.2 Evolution of flowering plants

Before we get into the specifics of pollination, I would like to provide a brief overview of the evolutionary origins of plant–pollinator relationships. Flowering plants that produce seeds encased in a protective ovary that develops into a fruit are called "Angiosperms." Apples, watermelon, and sunflowers are angiosperms because their seeds are encased in tissue that once was the ovary of a flower. According to the fossil record, Angiosperms first appeared about 135 million years ago during the Early Cretaceous period. By the Late Cretaceous period, there was greater species diversity among Angiosperms than any other plant group (see refs [2,3]). Today, in all but a few terrestrial locations, Angiosperms comprise at least 90% of the plant species, even in coniferous (i.e., pine) forests [4]. Many factors contributed to the success of Angiosperms, but a key factor is the evolution of relationships with pollinators. Birds, bats, and insects transferred pollen among plants as they foraged for nectar and pollen, causing flowers to set hybrid seed. Seed dispersal mechanisms such as barbed seed coverings that stick to animal fur and fruits that are food for animals spread seeds to new locations. Due to pollen transfer between plants (i.e., outcrossing), some of the hybrid seeds could germinate and establish stands under conditions and in sites where they did not previously exist.

3.3 Plant reproduction

Plants can propagate in two ways: sexually and asexually. Asexual reproduction is when a new plant is grown from part of a parent plant. Plants produced asexually are clones of the parent plant and genetically similar. The means by which plants reproduce asexually depend on the species. Some plants like strawberries produce stems called runners or stolons that grow horizontally and can produce buds that develop into new plants. New plants or even whole trees also can emerge from underground root systems. Entire groves of aspen trees can arise from a single tree and are clones connected by a common root system (Fig. 3.3).

Plants also reproduce sexually. Flower formation is needed for sexual reproduction because flowers house the plant's reproductive organs. Sexual reproduction occurs when pollen from the male part of a plant (anthers) is transferred to the female flower parts (pistil) (Fig. 3.4). Sometimes pollen is transferred within the same flower or among flowers on the same plant. This is self-pollination. In other instances, pollen is transferred to a flower on a different plant of the same species. This is cross-pollination. The arrangement of male and female flower parts and the timing of when those parts mature can largely determine whether foraging visits by pollinators result is self- or cross-pollination.

FIGURE 3.3 **Grove of aspen trees (*Populus tremuloides*) as an example of vegetative reproduction in plants.** Grove of quaking aspen trees (*Populus tremuloides*) in the White Mountains of eastern Arizona, USA. Although the grove looks like it is made of individual trees, the trees sprout from a common root system and are genetically similar. The sprouting of aspen trees from a common root system is an example of vegetative reproduction in plants. *Photograph by Gloria DeGrandi-Hoffman.*

FIGURE 3.4 **Reproductive parts of a flower from a walking stick cactus (*Cylindropuntia spinosior*).** Reproductive parts of a flower from a walking stick cactus (*Cylindropuntia spinosior*). The pistil is the female part and the stamens are the male part. Pollen grains are shed from the anthers. When pollen is deposited on the stigma, the grains germinate and grow down the style and fertilize the ovules in the ovary located at the base of the style. The fertilized ovules develop into seeds. Notice the pollen grains on the honey bee collecting nectar from the flower.
Emily Watkins deJong.

Most plants produce flowers during specific times of year. When a plant will flower depends on internal and external cues that affect a particular type of plant cell called a meristematic cell. Meristematic cells are comparable to stem cells in animals and can develop into a range of different plant tissues depending on where the cells are located. Meristematic cells occur in shoots, buds, root tips, and other parts of a plant where growth occurs. When a stem is cut from a plant and put in water, the roots that grow from the base of the stem are because of the meristematic cells located there.

The location of meristematic cells determines the type of tissue they form. If the cells are at the tips of a vegetative shoot, they develop into leaves, branches, and stems. If the shoot is reproductive, the meristematic cells become flowers. Flowers develop at the tip of reproductive shoots in concentric rings called whorls. Many types of flowers have four whorls. The two outer whorls make up the sepals and flower petals (i.e., the perianth) (Fig. 3.4). These whorls are seen when flowers open on plants like Rose of Sharon and other *Hibiscus* species. The two inner whorls become reproductive organs.

The first inner whorl develops into the stamens and anthers (male reproductive structures) and the inner most fourth whirl develops into the female plant parts: the pistils and ovules. At the tip of the pistil is the stigma. The surface of the stigma is moist and sticky. Pollen grains that land on the stigma germinate and send pollen tubes down the style of the pistil and to the ovaries where ovule fertilization and seed set occur. As the seeds mature, the ovarian tissue around the seeds swells to encase the seeds. Sometimes, the casing is woody with spines (e.g., *Datura* spp.). In other plants, the casing around the seeds becomes soft and sugary as it matures, as occurs in apples (*Malus* spp.) and cherries (*Prunus* spp.).

3.4 Pollination

When we look at flowers, we see beautiful colors and stunning shapes and smell fragrant odors. However, for honey bees and other pollinators, a flower is a communication agent that is seeking to build a relationship with the pollinator. The communication between flowers and pollinators and the relationships that have emerged are so successful that almost 80% of flowering plants either rely on or benefit from pollinators for seed set [5]. However, establishing the relationship is not as easy as it seems. Many plants flower for only 2–3 weeks in the spring or summer, so time for attracting pollinators is limited. There also can be many plant species blooming simultaneously, causing competition for pollinators' attention. Plants also have to be selective in the pollinators they attract to their flowers. Flowers need to attract individuals that are physically capable of moving pollen from male to female flower parts.

Plant communication with pollinators is like establishing a dialog between persons that speak different languages. The way to start is by looking for common ground. Flowers can find common ground with potential pollinators by producing odors or scents that pollinators can detect. Floral scents can send signals to pollinators from greater distances than other cues such as floral colors and are probably the first line of communication with potential pollinators. For the scent to attract a pollinator, however, there needs to be a physiological connection between the odor and the prospective pollinator. Specifically, bees need appropriate neural receptors where the scent molecules can bind and send signals to the brain that associate the odor with a food reward. The association will trigger a behavioral response (i.e., attraction to the flower). In plants pollinated by honey bees, the scents emanating from flowers can be the same as those produced by nestmates in the hive. For example, a volatile compound called E-β-ocimene is emitted from the flowers of many plant species [6]. Honey bees are familiar with the odor of E-β-ocimene and have neural receptors for it because queens produce it to signal that she is mated and laying eggs [7]. Larvae also produce E-β-ocimene as a signal that they need to be fed [8]. Other plants such as buckwheat (*Fagopyrum esculentum*) and Mexican sunflower (*Tithonia diversifolia*) communicate with their pollinators by producing flowers that secrete nectar with components of queen mandibular pheromone [9]. Bees detect the pheromone because it is secreted by queens and stimulates worker behaviors in the hive such as forming a retinue around the queen and grooming and feeding her [10].

Pollinators detect blooming plants due to the odors they produce, but it is the distinctive colors of petals that bees use to differentiate flowers from leaves. Honey bees see flowers very differently than we do. Humans have two types of photoreceptors; ones that can detect objects in dim light (rods), and others (cones) that enable us to see red, green and blue. Honey bee color vision is based on three types of photoreceptors that see ultraviolet, blue, and green regions of the color spectrum [11]. Plants tailor their flowers to the visual capabilities of honey bees. There are distinctive colors and patterns that honey bees and other pollinators can distinguish and learn to associate with nectar and pollen rewards (see refs [11,12]).

Flowers produce odors and colors to attract pollinators, but like all good relationships, there need to be mutual boundaries. Plants have to attract foragers to flowers but need to prevent visits that do not lead to pollination. Honey bees need to find flowers that provide rewards (especially nectar) containing more calories than they are expending during foraging. Systems are in place to meet both plant and pollinator needs. From the plant's perspective, the positioning of the reproductive structures within the flower (i.e., anthers, pistils, and stigma) can increase the chances that honey bees pick up pollen on their bodies while foraging and transfer it to female parts of the flower.

From the perspective of the honey bee, the most attractive plants are those with large numbers of rewarding flowers in a relatively small area. Ideally, foragers can collect rewards by walking among flowers rather than flying between plants. The large numbers of blossoms on fruit or nut trees such as almonds, apples or cherries makes them highly attractive to pollinators. Each blossom provides small amounts of nectar, but foragers can visit multiple blossoms without leaving a single branch, and collect full nectar and pollen loads by expending minimal energy (Fig. 3.5).

In addition to boundaries, plant–pollinator relationships also have a tendency toward manipulation, especially in the composition and distribution of nectar. Nectar is primarily sugars (glucose, fructose, and sucrose) and water but there also are amino acids, proteins, organic acids, and lipids that can comprise essential oils. Inorganic ions (e.g., sodium, potassium, and mineral salts) also may be present. Some compounds in nectar are produced by the plant and others might be produced by microbes living in the nectar (see ref. [13]). Though nectar provides food to foragers that followed the cues and find the flowers, other componds in nectar may have other functions.

Nectar rewards reinforce the pollinator's response to floral cues (i.e., collecting nectar and pollen) and increase the chances that the pollinator will return. The relationship between flowers and pollinators, however, might not

FIGURE 3.5 **Blossom clusters on an almond (*Prunus dulcis*) tree.** Honey bees foraging on almond (*Prunus dulcis*) blossoms. The blossoms are densely packed on the limb so that bees can walk among flowers and collect nectar and pollen. The floral arrangement is highly attractive to honey bees and other pollinators.
Photograph by Emily Watkins deJong.

be that simple. Recently, compounds have been detected in nectar that are not food sources for pollinators. The compounds can be nectar toxins that deter pollinator species that might visit but not pollinate the flower [14]. There also may be secondary metabolites in nectar that have pharmacological effects on the foragers. For example, nectar can contain caffeine. This compound can enhance foragers' memory of the reward so that they are more likely to associate it with the floral scent [15]. In a study where caffeine was added at field–realistic levels to sucrose solutions, it caused honey bees to increase their recruitment of new foragers. Once recruited, caffeine increased the frequency and persistence of foragers to the location [16]. Another compound in nectar that has profound effects on learning is nicotine [17]. In studies with bumblebees, nicotine enhanced bees' ability to learn colors and to continue foraging flowers containing nectar with nicotine even when they had less reward than those without the compound.

Plants can manipulate foraging behaviors in other ways such as how nectar is distributed among flowers. Not every flower needs to contain nectar for a plant to be attractive and maintain the pollinator's attention. Instead, flowers can provide intermittent rewards and still benefit from the constancy individual honey bees exhibit to flower patches [18]. However, intermittent reinforcement canot be carried too far. The dance language honey bees use to communicate locations of nectar sources to their nestmates may thwart the strategy. In addition to directions, the dances contain information on the distribution of rewards and their frequency. Patches of flowers with low frequency of rewards have little or no foragers communicating their location [19]. So, it is a balancing act for plants to expend as little energy as possible producing nectar while still producing enough for flowers to attract pollinators.

While the primary goal for the plant is to attract pollinators, it is as important that the pollinators cross-pollinate the flowers so that hybrid seed is produced. Attracting numerous pollinators can facilitate pollen transfer between plants (i.e., cross-pollination) because if foragers collect all the nectar and pollen from a flower, and the resources are not replaced by the plant, there ultimately will be primarily nonrewarding flowers. Pollinators that consistently encounter nonrewarding flowers are likely to switch to another plant where the chances of finding rewarding flowers might be higher. In the process of moving between plants, foragers can cross-pollinate the first few blossoms they visit [20]. However, honey bees moving from plant to plant cannot account for all the seed that is produced. This suggests that other systems are in place for cross-pollinating blossoms that incorporate the foraging population on a colony-wide basis.

3.5 Cross-pollination systems

At first glance, the foraging behavior of honey bees appears to run counter to cross-pollinating flowers. Honey bees are "flower-constant" during a foraging trip, meaning that they remain on a single flower species as long as it offers nectar or pollen [21]. However, two concurrent systems, one belonging to the plant and the other to the honey bee, are interacting to increase the likelihood of cross-pollination and production of outcrossed (hybrid) seed.

Plants can increase the chances of setting hybrid seed in several ways [22]. There are genetic systems that prevent ovule fertilization with self-pollen. Commonly referred to as self-incompatibility, ovule fertilization and seed set with pollen from the same flower or plant are prevented due to interactions between proteins on the pollen grain and those in the style or ovary of the pistil. In self-incompatible plants, if pollen from the same flower or plant lands on the stigma, it may germinate but pollen tube growth to the ovaries and fertilization of ovules will not occur. Some examples of self-incompatible plants are those in the mustard Family (Brassicaceae), and most apple, almond, and coffee varieties.

Another way plants can increase outcrossing and hybrid seed set is to have reproductive parts of flowers develop and mature at different times. For example, the male part (anthers) of each sunflower floret will mature and shed pollen before the female parts (pistil and stigma) open. Outcrossing also can be encouraged if pollen is shed before the pistil and stigma are receptive.

From the perspective of the honey bee, foragers strive for maximum efficiency when collecting nectar and pollen. Foragers prefer to remain on the same plant and collect nectar and pollen from multiple flowers, ideally within walking distance of each other. Foraging confined to flowers on the same plant facilitates self-pollination. So how do honey bees cross-pollinate flowers?

Fruit trees are a good system for studying how honey bees can cross-pollinate flowers because most varieties of apples and many other fruit trees exhibit self-incompatibility. Pollen that is transferred among blossoms on the same tree will not fertilize ovules and set seed (or apples). In fruit trees like apple, self-incompatibility extends to all trees of the same variety because commercially produced trees are propagated vegetatively. A cutting from a tree of a particular variety is grafted on to a rootstock and a new tree of that variety is produced. This means that trees of the same variety are genetically similar. Moving pollen between trees of the same variety is essentially self-pollination. For fruit to set, apple blossoms must be cross-pollinated with pollen from a different apple variety. For a long time, it was assumed that all the fruit you see on an apple tree was the result of bees moving between trees of different varieties while collecting nectar and pollen.

So, the question is, are all the apples on a tree the result of cross-pollination from honey bees moving between trees of different apple varieties? Studies of honey bees foraging in apple orchards indicated that bees confine their foraging flights to a single tree or row and rarely cross between rows [23]. If foragers visited only one tree or row of trees, they should have pollen from mostly one apple variety on their bodies, and fruit set should be an infrequent event. However, fruit set is not a rare event. Apple trees usually set so much fruit that some needs to be removed so that the apples grow to the size we see in fruit stands and grocery stores. However, orchards in bloom can attract thousands of bees. Maybe enough bees move between trees, especially if they cannot find blossoms with nectar, to account for all the fruit set.

The possible scenarios that would account for the amount of fruit that is set in an apple orchard are difficult to test in the field. Many variables come into play that affect bloom and foraging activity such as weather conditions, the number, arrangement, and varieties of trees in the orchard, and the size of the foraging population of honey bees. Testing scenarios with so many uncontrolled variables is best accomplished by building models of the system.

A cross-pollination and fruit set model for apples was constructed as a computer simulation model that describes the system mathematically. The model includes information on weather conditions, specifics of the orchard layout, and size of the honey bee foraging population [24]. Using hourly temperatures, the model tracks the number of open blossoms on trees of each variety throughout bloom. Warmer temperatures cause more blossoms to open, but they also age to petal fall sooner than when temperatures are cooler. Numbers of colonies in the orchard along with hourly temperature, wind speed, light intensity (i.e., cloud cover), and rain are used to predict the number of foraging bees on the apple trees. By combining predictions on the foraging population size with the number of flowers bees visit on a foraging trip to collect full nectar or pollen loads, the model predicts the number of open blossoms that were visited by foragers and those that were not visited and could still be rewarding. Using estimates of the number of nonrewarding (i.e., previously visited) blossoms and those that were not visited, probabilities are calculated for the chances of foraging bees consistently finding previously visited unrewarding blossoms and moving to another tree. Depending on the orchard design, bees could move to trees of the same cultivar (scored as no fruit set), or to trees of a different variety (scored as fruit set).

The fruit set predictions from the model were compared with actual fruit set in orchards of the same design and colony numbers and under the same weather conditions. Actual fruit set was many times higher than that predicted by the model, indicating that cross-pollination was occurring by means other than bees moving between different varieties of trees. The model predictions also pointed to the possibility that the population of foragers that had cross-pollen on their bodies was much larger than expected from bees moving between trees of different varieties.

So, how large is the population of honey bees with cross-pollen on their bodies while foraging apple trees? The cross-pollinating population on apple trees was estimated by removing pollen from the bodies of honey bees foraging apple blossoms. Pollen from different apple varieties and from other plants in bloom in orchards was identified with scanning electron microscopy. The pollen analysis showed that honey bees foraging on apple blossoms had pollen from many apple varieties on their bodies as well as pollen from other species of plants that were blooming in and around the orchard [25].

So, how does a honey bee that returns to the same tree or row of trees during a foraging day (see ref. [23]) get pollen on its body from other apple varieties and plant species? Remember that the foraging population in a hive is spread out among various flowering plants and pollen from many plant species is coming into the hive when favorable foraging conditions occur. The mixtures of pollen from different plants helps ensure that the colony meets its nutritional requirements.

In the hive, bees are in constant physical contact with each other as they walk over and next to nestmates making their way across the combs. Foragers contact their nestmates when they enter the hive and pass by guard bees. The foragers with nectar are contacted by nestmates that receive their nectar loads. Foragers with pollen can contact nestmates as they go to the combs where pollen is stored. During physical contacts, pollen can be transferred among bees so that all adult workers have some pollen on their bodies whether or not they ever foraged on flowers.

Although it is easy to imagine that bees can pick up pollen from their nestmates, the next question is do they pick up enough viable pollen to cross-pollinate flowers? This question was answered with a study where newly emerged adult worker bees that had no pollen on their bodies were marked and released in hives located in apple orchards that were in bloom. After several hours in the hive, the marked bees were collected and used to pollinate apple blossoms. Mesh bags were put over the blossoms so that other bees could not visit them, and that the only pollen deposited on their stigmas was from the marked bee. The marked bees were able to cross-pollinate the blossoms and cause fruit set [26]. Drones also were collected from the hives. Even though they do not forage on, the drones had pollen on their bodies and could set fruit.

The ability of honey bees to obtain pollen in the hive from contacts with nestmates makes honey bees highly efficient cross-pollinating agents. In agricultural systems with crops like apples that require cross-pollination for fruit set, transferring pollen in the hive enables honey bees to generate a foraging population with high cross-pollinating potential, especially when multiple varieties of trees are blooming in synchrony and are attractive to foraging bees. In undisturbed ecosystems, foraging honey bees can generate high rates of cross-pollination and hybrid seed production that is essential for maintaining genetic diversity within species.

Pollen transfer among workers in the hive can create a large population of cross-pollinating bees, but there are constraints. In apple orchards with varieties that are self-incompatible, at least two varieties need to be in bloom simultaneously for cross-pollen to be available. Parts of the foraging population in a colony must be visiting each of the varieties so that pollen from different varieties enters the hive and is transferred among nestmates. Bees also need to be collecting large amounts of pollen from different varieties for it to be transferred in the hive among nestmates. The size of a colony's foraging population and the amount of pollen collected depend on colony size. Large colonies in orchards with multiple varieties in bloom can send out foraging populations that cross-pollinate the first few blossoms they visit on each foraging trip.

Pollen transfer in the hive is not the only way flowers can be cross-pollinated by honey bees. Pollen is sticky and often is transferred from the anthers to the stigma in clumps of pollen grains. Other bees can pick up the clumps of pollen from the stigma when foraging flowers and spread to other flowers they visit. Whether seed can be set by bees spreading pollen across flowers was tested in a study with commercial sunflower varieties [27]. To control the genetics of commercial sunflower varieties, plant breeders produce a genetic line for the variety that does not produce pollen (i.e., "male-sterile" lines). The male-sterile line is planted with genetic lines that produce pollen (i.e., male-fertile). Seed set occurs in male-sterile varieties only when pollen from male-fertile lines is deposited on the stigma of male-sterile flowers. In the study, male-sterile and male-fertile sunflower lines were planted as they are in commercial fields where sunflowers are grown for seed. The varieties were planted in solid rows of either male-sterile or male-fertile plants. Both varieties are very attractive to honey bees and wild bees. Bees collect nectar from male-sterile flowers and pollen and nectar from male-fertiles.

The study showed that although honey bees were foraging on male-sterile sunflowers that were not producing pollen, nearly all the foragers had sunflower pollen on their bodies [27]. The foragers might have picked up the pollen from nestmates in the hive as they do in apple orchards. When the amount of pollen bees obtained from nestmates in the hive was measured, clearly some pollen could be acquired there. However, the amounts of pollen bees picked up in the hive were much lower than that found on honey bees foraging on male-sterile sunflowers even if the plants were four to eight rows away from pollen producing male-fertile plants.

A second study was conducted focusing on factors that might account for the amount of pollen on the bodies of honey bees foraging male-sterile sunflowers [28]. The study revealed significant positive relationships between the amount of pollen on the bodies of honey bees and the number of wild (i.e., non-*Apis*) bees foraging male-sterile flowers. As wild bees increased on male-sterile sunflowers, so did the amount of pollen on the bodies of honey bees foraging in male-sterile rows [28]. Wild bees collect pollen and nectar on foraging trips and may have visited male-fertile sunflowers first for pollen and then moved to male-steriles. The clumps of pollen deposited on male-sterile florets might then be spread across the sunflower florets by honey bees foraging for nectar on male-sterile sunflowers. The study also showed the importance of having both honey bees and wild bees foraging on hybrid sunflowers, as together they might maximize seed set on the male-sterile lines.

Both in-hive pollen transfer and the movement of previously deposited pollen among other flowers probably increase hybrid seed production in many plant species. In-hive pollen transfer was found to occur in the pollination of almonds and probably contributes to outcrossing in other crops and plants in undisturbed habitats where honey bees spread their foraging population across different varieties and flower patches.

3.6 Climate change effects on plant attractiveness, pollination, and fruit set

Since flowering plants evolved about 135 million years ago, mutually beneficial relationships have been established between plants and their pollinators. Those relationships have supported and sustained both groups. However, the relationship is being threatened in many ways by climate change. Increasing incidences of floods and fires certainly have devastating effects on plant and pollinator communities. However, heat and drought also impact each component in the system that evolved to attract honey bees and other pollinators to flowering plants and provide nutrients for their health and population growth. Plants also are affected. Although pollination might occur, high heat and water stress can disrupt the system that includes pollen tube growth, ovule fertilization, and seed set [29].

Perhaps the most noticeable effects of climate change, particularly higher temperatures, are the effects on the time when plants flower. Some plants that experience high temperatures may flower earlier in the season and bloom before their pollinators are available [30]. Other plants might be less likely to flower due to heat stress or produce fewer and smaller flowers (see ref. [31]). The flowering period also can be shortened, since the length of time between flowers opening and shedding their petals and senescing (i.e., bloom period) is affected by temperature. The ability of the bloom to attract pollinators also can be impacted by high temperatures. Flowering plants attract pollinators from long distances using floral scents comprised of volatile compounds that are temperature-sensitive [32,33]. Studies measuring effects of temperatures on floral odors show a threefold decrease in floral scents from buckwheat plants grown at high temperatures and a loss of detectable odors in strawberry flowers [32].

In addition to heat, the effects of climate change on rainfall patterns and increased frequency and duration of drought also are disrupting plant–pollinator interactions and ultimately seed set. Plants normally experience periods of water stress causing wilting that is alleviated when water is added to the soil. Drought differs from simple water stress. During droughts, water availability is significantly below normal levels for a specific site or locality for an extended period. Plant responses to drought often are seen in flower production. Flowers are energetically expensive to plants when considering the carbon, minerals, and water needed to produce floral tissue with colorful pigmentation, volatile chemicals that generate odors to attract pollinators and nectar and pollen rewards. Flowers also are a major contributor to water loss in plants (see ref. [34]). During droughts, plants produce fewer and smaller flowers or none at all.

Similar to heat stress, drought conditions decrease the length of the flowering period, the number, size, and color of flowers, and the amount of nectar and pollen they produce [35]. Since nectar is mostly water, under drought conditions, flowers secrete less nectar with more concentrated sugar levels [34]. Nectar with high sugar levels is thick and syrupy and difficult for honey bees to collect through their narrow mouth parts (proboscis). Pollen production and possibly nutrients in the pollen also can be reduced when plants are water stressed.

For plants, the purpose of their relationship with pollinators is seed set. High temperatures, even for brief periods, can impact the development of pollen grains, pollen germination, pollen tube growth, ovule fertilization, and seed/fruit set. Crops that bloom in the spring might be the most affected because they have greater sensitivity to heat extremes than those that bloom in summer [36]. For example, blueberry plants that flower in May have lower fruit set when temperatures exceed 95°F (35°C) because pollen germination, pollen tube growth, and subsequent ovule fertilization and seed set are significantly reduced [37]. In undisturbed ecosystems, the distribution of plant species and composition of plant communities are driven by temperature effects on pollen germination and ovule fertilization [38]. With higher temperatures, regions where plants can currently establish might shift due to an inability to set seed even if pollination occurs.

The effects of climate change on pollination have far reaching consequences on ecosystem stability, food security, and human health. In undisturbed ecosystems, poor pollination and reductions in seed set limits subsequent expansion of areas covered with plants. Those plants play key roles in soil stability, moisture retention, moderation of temperature, and cover and food sources for wildlife including the myriads of subterranean organisms and microbes occupying ecosystems beneath our feet. The effects of climate change on pollination of agricultural crops can have serious effects on food security and human health that are already being revealed. Reduced yields of honey bee pollinated crops impact more than 700 million undernourished people worldwide and 2 billion people globally suffering from micronutrient deficiencies [39].

The importance of honey bee pollinated crops and their connection to human health cannot be overstated. Although crops relying on honey bee pollination comprise a relatively small fraction of agricultural production when compared with volumes of wheat, corn, or rice, fruits crops, vegetables, nuts, and legumes supply the majority of micronutrients and dietary fiber required for optimum health, reduced cardiovascular disease and cancer prevention.

Human epidemiological studies show that greater consumption of fruits, vegetables, nuts, and legumes provides protection against a range of chronic noncommunicable diseases [39] and contribute to weight loss [40]. The effects that lower yields of honey bee pollinated crops have on the human diet and health are not evenly distributed across societies. Lower yields drive up costs and limit the availability of fruits and vegetables. Lower-income groups are more greatly affected, further widening the inequality of diet and health outcomes among socioeconomic groups. These topics will be discussed in greater detail in subsequent chapters.

References

[1] Klein AM, Vaissière BE, Cane JH, Steffan-Dewenter I, Cunningham SA, Kremen C, et al. Importance of pollinators in changing landscapes for world crops. Proc Biol Sci 2007;274(1608):303–13.

[2] van der Kooi CJ, Ollerton J. The origins of flowering plants and pollinators. Science 2020;368(6497):1306–8.

[3] Benton MJ, Wilf P, Sauquet H. The Angiosperm Terrestrial Revolution and the origins of modern biodiversity. N Phytologist 2022;233(5):2017–35.

[4] Kreft H, Jetz W. Global patterns and determinants of vascular plant diversity. Proc Natl Acad Sci U S A 2007;104:5925–30.

[5] Rodger JG, Bennett JM, Razanajatovo M, Knight TM, van Kleunen M, Ashman TL, et al. Widespread vulnerability of flowering plant seed production to pollinator declines. Sci Adv 2021;7(42):eabd3524.

[6] Farré-Armengol G, Filella I, Llusià J, Peñuelas J. β-Ocimene, a key floral and foliar volatile involved in multiple interactions between plants and other organisms. Molecules 2017;22(7):1148.

[7] Gilley DC, DeGrandi-Hoffman G, Hooper JE. Volatile compounds emitted by live European honey bee (Apis mellifera L.) queens. J Insect Physiol 2006;52(5):520–7.

[8] Maisonnasse A, Lenoir J-C, Beslay D, Crauser D, Le Conte Y. E-b-Ocimene, a volatile brood pheromone involved in social regulation in the honey bee colony (Apis mellifera). PLoS ONE 2010;5(10):e13531.

[9] Liu F, Gao J, Di N, Adler LS. Nectar attracts foraging honey bees with components of their queen pheromones. J Chem Ecol 2015;41:1028–36.

[10] Slessor KN, Kaminski L-A, King GGS, Borden JH, Winston ML. Semiochemical basis of the retinue response to queen honey bees. Nature 1988;332:354–6.

[11] Avarguès-Weber A, Dyer AG, Combe M, Giurfa M. Simultaneous mastering of two abstract concepts by the miniature brain of bees. Proc Natl Acad Sci U S A 2012;109(19):7481–6.

[12] Petrikin J, Wells H. Honey bee (Apis mellifera) use of flower pigment patterns in making foraging choices. J Kans Entomol Soc 1995;68(4):377–87.

[13] Barberis M, Nepi M, Galloni M. Floral nectar: fifty years of new ecological perspectives beyond pollinator reward. Perspect Plant Ecol Evolut Syst 2024;62:125764.

[14] Barlow SE, Wright GA, Ma C, Barberis M, Farrell IW, Marr EC, et al. Distasteful nectar deters floral robbery. Curr Biol 2017;27(16):2552–8.

[15] Wright GA, Baker DD, Palmer MJ, Stabler D, Mustard JA, Power EF, et al. Caffeine in floral nectar enhances a pollinator's memory of reward. Science 2013;339(6124):1202–4.

[16] Couvillon MJ, Al Toufailia H, Butterfield TM, Schrell F, Ratnieks FL, Schürch R. Caffeinated forage tricks honeybees into increasing foraging and recruitment behaviors. Curr Biol 2015;25(21):2815–8.

[17] Baracchi D, Marples A, Jenkins AJ, Leitch AR, Chittka L. Nicotine in floral nectar pharmacologically influences bumblebee learning of floral features. Sci Rep 2017;7(1):1951.

[18] Wells H, Wells PH. Honey bee foraging ecology: optimal diet, minimal uncertainty or individual constancy? J Anim Ecol 1983;52:829–36.

[19] Seefeldt S, DeMarco RJ. The response of the honeybee dance to uncertain rewards. J Exp Biol 2008;211:3392–400.

[20] Bailey SF, Hargreaves AL, Hechtenthal SD, Laird RA, Latty TM, Reid TG, et al. Empty flowers as a pollination-enhancement strategy. Evolut Ecol Res 2007;9(8):1245–62.

[21] Grant V. The flower constancy of bees. Bot Rev 1950;16(7):378–98.

[22] Jesson L. Reproductive strategies. Encyclopedia of applied plant sciences Vol. 2. Elsevier; 2017.

[23] Free JB. The foraging areas of honeybees in an orchard of standard apple trees. J Appl Ecol 1966;3:261–8.

[24] DeGrandi-Hoffman G, Hoopingarner R, Pulcer R. REDAPOL: pollination and fruit-set prediction model for 'Delicious' apples. Environ Entomol 1987;16(2):309–18.

[25] DeGrandi-Hoffman G, Hoopingarner RA, Baker KK. Identification and distribution of cross-pollinating honey-bees (Hymenoptera: Apidae) in apple orchards. Environ Entomol 1984;13(3):357–764.
[26] DeGrandi-Hoffman G, Hoopingarner RA, Klomparens K. The influence of honey-bee (Hymenoptera: Apidae) in-hive pollen transfer on cross- pollination and fruit set in apple. Environ Entomol 1986;15:723–5.
[27] DeGrandi-Hoffman G, Martin JH. The size and distribution of the honey bee (Apis mellifera L.) cross-pollinating population on male-sterile sunflowers (Helianthus annuus L.). J Apic Res 1993;32(3–4):135–42.
[28] DeGrandi-Hoffman G, Watkins JC. The foraging activity of honey bees Apis mellifera and non-Apis bees on hybrid sunflowers (Helianthus annuus) and its influence on cross-pollination and seed set. J Apic Res 2000;39(1–2):37–45.
[29] Rosenberger NM, Hemberger JA, Williams NM. Heatwaves exacerbate pollen limitation through reductions in pollen production and pollen vigour. AoB Plants 2024;16(5):plae045.
[30] Kudo G, Ida TY. Early onset of spring increases the phenological mismatch between plants and pollinators. Ecology 2013;94(10):2311–20.
[31] Scaven VL, Rafferty NE. Physiological effects of climate warming on flowering plants and insect pollinators and potential consequences for their interactions. Curr Zool 2013;59(3):418–26.
[32] Cordeiro GD, Dötterl S. Floral scents in bee-pollinated buckwheat and oilseed rape under a global warming scenario. Insects 2023;14(3):242.
[33] Cordeiro GD, Dötterl S. Global warming impairs the olfactory floral signaling in strawberry. BMC Plant Biol 2023;23(1):549.
[34] Kuppler J, Kotowska MM. A meta-analysis of responses in floral traits and flower–visitor interactions to water deficit. Glob Change Biol 2021;27(13):3095–108.
[35] Descamps C, Quinet M, Jacquemart AL. The effects of drought on plant–pollinator interactions: what to expect? Environ Exp Bot 2021;182:104297.
[36] Hedhly A, Hormaza JI, Herrero M. Global warming and sexual plant reproduction. Trends Plant Sci 2009;14(1):30–6.
[37] Walters J, Isaacs R. Pollen germination and tube growth in northern highbush blueberry are inhibited by extreme heat. HortScience 2023;58(6):635–42.
[38] Rosbakh S, Poschlod P. Minimal temperature of pollen germination controls species distribution along a temperature gradient. Ann Bot 2016;117(7):1111–20.
[39] Smith MR, Mueller ND, Springmann M, Sulser TB, Garibaldi LA, Gerber J, et al. Pollinator deficits, food consumption, and consequences for human health: A modeling study. Environ Health Perspect 2022;130(12):127003.
[40] Dreher ML, Ford NA. A comprehensive critical assessment of increased fruit and vegetable intake on weight loss in women. Nutrients 2020;12:1919.

Chapter 4

Colony as an organismal system that responds to the environment

The key to understanding honey bee colonies is to hold two thoughts in your mind simultaneously. The first is understanding the honey bee, especially workers, as individuals that have dynamic subsystems that can respond to changing environmental conditions. The second is understanding the colony as a whole system or organism with numerous interacting subsystems. The subsystems in a colony are comprised of coordinated groups of workers responding to their environment. Linkages among the subsystems generate a colony-level response.

In biology, an organism is defined as a living being capable of growth, reproduction, and responding to its environment. Organisms are capable of complex activities because their cells respond to changes in the environment. In honey bee colonies, worker bees are similar to cells in an organism. Worker physiologies and behaviors can change in response to their environment. The responses shared among individual workers are then manifested at the colony level. Food storage, comb building, nest defense, and swarming are examples of individual behaviors that collectively generate a colony-level (i.e., organism-level) response.

In this chapter, I will discuss colony-level responses to environmental factors most closely tied to climate change. These are nutritional stress, extreme heat, and cold. Climate change causes nutritional stress because heat waves and drought limit flowering and the nectar and pollen available to bees. High temperatures also test the ability of colonies to maintain hive temperatures especially in ranges that are conducive to brood rearing. When warm temperatures persist beyond the summer and into the fall, preparations for overwintering by individual workers can be compromised, endangering the colony's ability to successfully overwinter. Subsystems in colonies that generate food gathering and distribution and that maintain colony temperatures are resilient and can adjust to periods of stress; but there are limits. Systems have boundaries within which they function. Climate change is challenging

The Honey Bee and Climate Change. DOI: https://doi.org/10.1016/B978-0-443-28995-8.00014-3

those boundaries, making it difficult for honey bees and other pollinators to survive in areas that were once favorable and where populations thrived.

4.1 Nutritional stress

Climate change creates nutritional stress for honey bees and other pollinators because of the effects that extremes in temperature and precipitation have on flowering plants. Some effects of climate change on colony nutrition are obvious and associated with extreme weather events such as flooding from hurricanes, typhoons, and atmospheric rivers. These events cause nutritional stress by reducing or eliminating flowering plants. Likewise, wildfires occurring from warm and dry conditions in spring and summer also create overt nutritional stress from the devastation of plant communities often on enormous scales. Like floods, landscape recovery after wildfires may take a season or more before pollinators can again be supported by flowering plants.

There also are more subtle effects of climate-related events such as heat waves or droughts that can reduce flower numbers and the nectar and pollen they produce. The consistency and nutrients in floral resources also can be altered. Colony responses to nutritional stress begin in worker bees and expand into a colony-wide response. In general, two colony subsystems adjust to nutritional stress: brood rearing and foraging. Both emerge from physiological subsystems in worker bees.

Throughout the day in a honey bee colony, groups of workers are performing diverse tasks. The tasks are divided among the workers largely based on their age and physiology. Young workers (1–3 weeks old) that consume pollen and have highly developed hypopharyngeal glands (HPGs) take care of brood. After about a week, the workers consume less pollen and the HPG are less active. The workers perform house duties such as storing nectar, converting it to honey, and removing debris and dead bees from the hive. When workers are older (i.e., more than about 4 weeks), they engage in nest defense at the hive entrance. As flight muscles mature and neurophysiological capabilities such as navigation and pattern recognition develop, workers transition into foragers [1].

Performing behaviors and tasks based on age generally occurs when colonies are not experiencing stress. Keeping adult workers within the safe confines of the hive for more than 3 weeks helps ensure that a worker's lifespan exceeds the time needed to replace them (i.e., 21 days), and the colony will grow. However, the ages when workers perform various tasks can be altered depending on colony and environmental conditions. This is part of the built-in flexibility of colonies that enables workers to obtain information on colony conditions, often through interactions with nestmates, and adapt their behaviors accordingly. One such adaptation is altering the age when workers begin to forage when a colony is experiencing nutritional stress.

The ability of workers to detect when their colony is experiencing a food shortage and shift the age when they first forage was revealed in a study where colonies were provided with either abundant or starvation-level amounts of honey. The colonies contained workers that emerged together as adults so that all the bees were 1-day old when the colony was established. In this way, the age of workers when they began foraging could be determined. Workers in colonies given starvation-level amounts of honey began foraging at younger ages than those in colonies with abundant honey [2]. When the starving colonies were fed, the number of new foragers declined in about a day. When colonies that were fed abundant amounts of honey were switched to starvation levels, the number of new foragers increased.

The relationship between available food and age of first foraging might be related to the amount of nutrients young bees store in their fat bodies. Young workers caring for brood and performing tasks inside the hive (i.e., house bees) have fat bodies with high protein and lipid levels. As the workers age, the levels decrease and are the lowest in foragers. When colonies are nutritionally stressed, stored nutrients in the fat body decline and this accelerates the transition from house bee to forager [3].

Workers that begin foraging at younger ages than expected (i.e., less than 21–28 days) are called "precocious foragers." Workers that forage earlier in life are not completely ready for the task. Precocious foragers have underdeveloped flight muscles and a greater tendency to drift to other colonies because of underdeveloped navigational skills. They also restrict their foraging areas to flower patches close to their hive. Most importantly, however, precocious foragers gather resources for fewer days and have shorter lifespans than workers that began foraging as older bees (see ref. [4]).

Precocious foraging is a response by individual workers to nutritional stress, but the effects are felt by the colony as a whole. This is because foraging earlier in adult life undermines the two interconnected components required for colony growth: the amount of brood reared and the longevity of worker bees [5,6]. When hive bees are shifted to the foraging population, there are fewer bees for tasks that support brood rearing. Workers that start foraging at a younger age also have a shorter lifespan [4]. If workers continue to forage at a younger age, a negative feedback loop forms. There will be fewer bees rearing less brood causing fewer new adult workers to emerge so less brood will be reared. If nutritional stress persists, the combination of reduced brood rearing and shorter lifespans in adult workers will cause a downward population spiral that will gain momentum over time. Colony loss might be averted, however, if a source of nectar and pollen comes into bloom or beekeepers provide food to the colony in the form of sugar syrup and pollen/protein supplementation. However, there can be long-term effects on colony survival, especially if the nutritional stress prevented populations from growing to sizes that enable successful overwintering.

If increasing the foraging population does not relieve nutritional stress, workers bees, especially those that care for brood, activate alternative options to conserve food resources. One such option is when nurse bees use nutrient reserves in their bodies to rear brood [7] . Another option is to reduce the number of feeding visits to larvae [8]. As a last resort, workers will stop rearing brood and begin cannibalizing eggs and larvae (see ref. [9]).

During nutritional stress, workers preferentially consume eggs and younger larvae [10]. Workers consume younger larvae perhaps because they are likely to be malnourished [8] with little chance of developing into healthy adults. Conversely, older larvae (4–5 days old) are not as affected by pollen scarcity as young larvae. In fact, older larvae appear to experience preferential treatment by nurse bees perhaps because they represent greater energy investment by the colony [11]. To secure the investment and conserve pollen, cells are capped earlier in nutritionally stressed colonies compared with colonies that are well fed [10].

Regulating the amount of brood reared in nutritionally stressed colonies might also have a chemical component . The volatile compound, E-β-ocimene, is released by eggs, young larvae, prepupae, and laying queens (see ref. [12]). Colonies that are deprived of pollen and are cannibalizing brood have low ocimene levels [12] perhaps because there are fewer eggs and young larvae that release ocimene as a signal that they need to be fed [13]. If ocimene is added to nutritionally stressed colonies, workers rear brood. It appears that increasing ocimene levels might be a cue that nutritional stress is lessening and brood rearing can resume [12]. Initially, small numbers of eggs are reared into larvae, but as more larvae are retained and ocimene levels rise, a positive feedback loop can be established that creates increasing numbers of eggs being reared, a stronger ocimene signal, and colony recovery from nutritional stress.

Recycling nutrients from cannibalism of young larvae and eggs is effective in preserving critical nutrients during periods when pollen is in short supply. In studies where nutrients from consumed larvae were chemically labeled, those nutrients were found in HPG secretions and in food exchanged among workers [14,15]. In fact, workers could rear brood successfully using only cannibalized larvae as a food source (see ref. [16]).

Precocious foraging and brood cannibalism can help colonies survive periods of food shortages. However, there can be long-term consequences and generational effects on honey bee populations across landscapes. The long-term effects include insufficient colony growth in the summer and possible colony loss over the winter. Generational effects of nutritional stress extend beyond the years when they occur because nutritional stress limits or curtails drone production. Only large healthy colonies with abundant resources can afford to rear drones. While drones do relatively little for the colony that rears them, they are vehicles for disseminating the colony's genes into the environment, and investments in the future of honey bee populations across ecosystems . In this chapter and in others, I discuss the importance of genetic diversity in a colony. As mentioned in Chapter 2, queens mate with multiple drones so that the colony has many patrilines (i.e., genetic diversity). Colonies with many patrilines usually are larger, forage more efficiently, and have less disease than those with fewer patrilines [17]. If queens are to mate with numerous drones from different colonies, there needs to be many robust colonies rearing drones. Regions where flowering plants are scarce because of fire, drought, flooding, or heat waves will have nutritionally stressed colonies and few drones. New queens in those areas might not mate with enough drones to create the genetic diversity needed for a healthy resilient colony. Negative feedback loops can emerge at colony and population levels from the conditions that create extended periods of nutritional stress. Colonies with poorly mated queens and low levels of genetic diversity in the worker populations may not reach sizes that are large enough to rear drones and therefore continue to create conditions where queens are poorly mated. Over time, honey bee colonies might be rare in those areas and limit pollination and seed set in plants they forage.

4.2 Thermoregulation

4.2.1 Cooling the hive

Cooling the hive during periods of extreme heat is essential for the survival of bees in all life stages. Adult bees can tolerate high temperatures for short periods, but developing brood are far less heat tolerant than adults. Optimum temperatures for brood development are 32°C–36°C (89.6°F–96.8°F) [18]. If brood nest temperatures reach 37°C (98.6°F), the emergence of adult bees from the pupal stage falls by 40% and at 38°C (100°F), no adultse merge [19]. Short-term exposures to high temperatures also have negative effects such as deformed wings, legs, and stingers [20]. Learning capacities also are negatively affected [21,22]. Drones exposed to high temperatures during development have reduced sperm viability [23].

Colonies contain 30,000 or more adults and developing brood and are warmed by the heat generated from metabolic activity. Add in ambient temperatures over 33°C (91°F), and hives can quickly become warm enough to cause mortality to developing brood and adults. To prevent overheating, honey bees have evolved multiple systems to regulate temperatures and maintain optimums for brood development and adult activities.

I live in Tucson, Arizona, and in the summer, afternoon temperatures can reach 43.3°C (110°F). The inside of a parked car can get so hot, it is difficult to sit on the seat or touch the steering wheel. Honey bee colonies also live in Tucson, so how do they deal with such high temperatures? Without a cooling system, the hive would be like the inside of a car on a hot day with temperatures beyond what a bee could survive.

In Tucson, our most popular means of cooling homes is an evaporative cooling system (i.e., swamp coolers). The system works by pulling in outside air and passing it over water-saturated pads. The heat in the air causes the water in the pads to evaporate. When water changes from liquid to vapor, it absorbs heat and cools the air surrounding it. The cool air created through water evaporation in the pads is blown into air ducts and circulates through vents in a home. The system works best when humidity is low. In Tucson, average humidity is little more than 20% in May and June, so evaporative cooling works well. As humidity rises with summer rains in July through September, however, evaporative cooling is no longer effective for keeping temperatures low.

Honey bees have an evaporative cooling system that is similar to the one we use in Tucson. Foraging bees collect water in their honey stomachs and take it back to the hive. The foragers regurgitate the water in the hive and spread droplets on their bodies and on brood cells. Workers also hold a thin film of water on their mouthparts. The water droplets evaporate in the hive with the aid of "fanner bees" (workers that fan the air with their wings) positioned on the combs and at the hive entrance. The fanning enhances evaporation and creates circulation of the cooler air [24,25]. The circulation has a pulse that resembles a typical breathing pattern with tidal air flow. Fanning causes air to move out of the colony entrance (exhale). There is a pause in the fanning and an influx of outside air into the hive (inhale), followed by a resumption of fanning [26].

If temperatures and humidity are too high for evaporative cooling alone, workers also can lower hive temperatures by clustering outside the hive at the entrance [Fig. 4.1]. Having fewer bees in the hive reduces the metabolic heat generated by the thousands of bees. Workers also can cool sections of the hive by forming insulating heat shields especially to protect the brood [27,28]. The shields are comprised of young workers that press the ventral side of their bodies against heated surfaces on the brood comb and absorb the heat. The workers then move to cooler areas of the hive and allow the heat to dissipate.

The tendency to participate in activities that cool the hive depends on the age and patriline of a worker. Colonies modulate their response to high temperatures by having some but not all workers participate in evaporative cooling. Colonies with diverse patrilines are able to regulate temperatures within a significantly narrower range around the optimum brood nest temperature (35°C, 95°F) compared with colonies having a single patriline. In genetically diverse colonies, workers that respond first are those in patrilines with the lowest tolerance for temperatures above 35°C . If temperatures remain high, workers in patrilines with higher heat tolerances participate in cooling activities. The genetic diversity in a colony and the thresholds different patrilines show in response to those temperatures generate a graded response to high temperature. In contrast, temperatures in experimental colonies comprised

FIGURE 4.1 **Honey bees clustered at the entrance of a hive to reduce heat during conditions of high ambient temperatures.** Clusters of honey bees at a hive entrance. The bees are at the entrance to reduce the temperatures inside the hive. This behavior called "bearding" occurs when ambient temperatures and humidity are too high for bees to cool the hive with evaporative cooling. *Photograph from Emily Watkins deJong.*

of workers with the same patriline fluctuated above and below 35°C without reaching a steady state [29].

Colonies maintain temperatures close to optimum primarily in the brood nest area. Temperatures are less tightly controlled on empty comb or where honey and pollen are stored. How much brood a colony can rear when temperatures are high depends on the area that bees can thermoregulate, and this depends on water availability and colony size. The interactions among colony size, thermoregulation, and brood area set up a feedback loop. There is a positive feedback loop for large colonies because they can thermoregulate and rear brood over large areas of comb and continue to grow. Conversely, small colonies thermoregulate and rear brood in smaller sections of comb. If less brood is reared, there will be fewer bees to thermoregulate the brood area. A negative feedback loop is established between the area where brood can be reared and the numbers of workers that can maintain hive temperatures and rear brood. If the heat persists, the negative feedback loop can cause populations to decline to a point where smaller colonies perish.

The effects of high temperatures can have both short- and long-term effects on colony growth and survival. In the short term, colonies might reduce brood rearing, especially if high temperatures limit the area in the hive that can be thermoregulated. There also might be a shortage of flowering plants and pollen, further curtailing brood rearing. In the long term, colonies might lose their queens, since under extreme high temperatures, mated queens can lose sperm viability [30]. If colony growth is limited by the size of the brood nest that can be thermoregulated and fertilized eggs laid by the queen, colonies might survive the summer and fall but not reach sizes where they can successfully overwinter.

4.2.2 Warming the hive

For colonies to remain active throughout the year, bees need to both cool in summer the hive and keep it warm during the winter. The system that generates heat to warm the hive has two main components. The first is food storage, especially honey. Colonies need sufficient amounts of honey to remain active while they are confined in the hive especially during the cold of winter. In temperate areas, colonies use 40–60 kg (88–132 lbs) of honey over the winter (see ref. [31]). The second component of warming the hive is the formation of the winter cluster.

The system that generates heat to warm the hive activates when ambient temperatures drop below 10°C (50°F). Bees crowd together particularly on the center combs of the brood nest and form a cluster. Although the cluster may look like a group of tightly packed bees, it has an architecture that is designed to preserve heat and prevent CO_2 from reaching toxic levels. At the core of the cluster is the queen. Bees at the core generate heat by their metabolic activity and by "shivering behavior." Shivering involves rapid contraction of flight muscles in the worker's thorax. Workers generating heat at the core of the cluster are called endothermic bees [32], and they can maintain temperatures ranging from 20°C to 34°C (68°F–93.2°F). However, endothermic activity is metabolically taxing so there is constant movement of endothermic bees into and out of the cluster core.

The winter cluster is in constant motion so that the heat generated by the bees is conserved. The cluster is warmest at the core that surrounds the queen. Bees are loosely assembled at the core to enable air flow. Cluster temperatures are cooler with increasing distance from the core, so the bees are more tightly packed together to preserve heat. Still, temperatures on the outer edges of the cluster can drop below 10°C (50°F) during the cold of winter. Bees become inactive at low temperatures and can fall from the outer edges of the cluster and die. To prevent this, workers on the outside of the core form a mantle around the cluster. The mantle previously was thought to insulate the cluster, but more recent findings indicate that the mantle acts as a heat sink [31]. The tightly packed bees on the mantle store rather than generate heat. When the heat in individual bees on the mantle dissipates, the bees move from the outer edge to the inner parts of the cluster. The cluster maintains its shape and organizational structure in the hive throughout the winter as the bees move from lower to upper regions of the combs. The movement is needed to ensure that the cluster remains in close proximity to combs containing honey.

Preparations for winter begin months in advance. The stores of honey needed for overwintering are collected in late summer and fall. Individual workers also prepare for overwintering by transitioning from short-lived summer to long-lived winter bees that will comprise the winter cluster. Maximizing the longevity of winter bees is essential for successful overwintering. The transition to winter bees occurs gradually during the fall and seems to be related to declining levels of brood rearing [33]. As colonies rear less brood, they begin storing resources. The first place resources are stored is in the bees themselves and then in comb cells. The food stored in winter bees results in greater dry weight, and higher amounts of total proteins, glucose, glycogen, and lipids than in summer bees [34,35]. There also are higher hemolymph protein levels, primarily vitellogenin [36,37], and lower juvenile hormone titers [38,39]. Winter bees also have lower amino acid levels in their hemolymph than summer bees, with the exception of those that enhance the ability of workers to generate heat. One such amino acid that is higher in winter bees is proline. Proline shuttles carbon from lipid reserves in the fat body to the flight muscles that generate heat for thermoregulation [40–42]. Trehalose, which enhances cold stress tolerance in insects, also is found at higher levels in winter bees compared with that in summer bees [40].

In addition to physiological differences between summer and winter bees, anatomical changes occur. The HPGs in winter bees can become smaller and the fat body larger than in summer bees [37]. While the fat body itself might be enlarged in winter bees, so are individual fat body cells, indicating increased metabolic activity.

Changes in fat body composition may contribute to winter bees living longer than summer bees. One important change is that fat body cells in winter bees show increased activity of an enzyme called telomerase [43]. Telomerase is an enzyme that binds to the ends of chromosomes and prevents the shortening of telomeres that are located there. The regulation of aging and lifespan in many organisms is associated with the length of telomeres [44]. Telomeres ensure proper chromosome function during cell division. With each round of cell division, telomeres shorten leading to cell senescence and aging. Higher telomerase activity slows the cell aging process by preserving the length of telomeres and this might help extend the life of winter bees.

When brood rearing ends in the fall, fat body lipid levels are higher and protein levels are lower than after overwintering when brood rearing resumes. Colonies start rearing brood while still confined in the hive in their winter cluster. The fat body lipids might be a primary source of nutrients for the brood, as there is a relationship between decreasing fat body lipid levels and increasing amounts of brood reared while the colony is confined in the hive [45]. The brood are critical to the colony's survival in the spring, as they will provide replacements for workers in the cluster that die once foraging resumes.

4.3 Overwintering and climate change

Overwintering has become more difficult, as indicated by consistently high colony losses. The losses may occur with greater frequency as warmer fall temperatures are expected to increase over expanding temperate zone regions [46,47]. Warmer falls and later winters affect the two components of successful overwintering: storing sufficient food and maximizing the lifespan of worker bees. Food stores are jeopardized because when fall temperatures are high, bees will forage. If flowering plants are not available, foraging is creating a caloric deficit for the colony just when food needs to be stored for the winter. The deficit is paid by the honey stored in the hive. Without sufficient honey stores, bees will be unable to thermoregulate the cluster and the colony can be lost from cold or starvation. Food limitations might also limit the amount of brood bees can rear while still confined in the hive. When foraging resumes in the spring, if there are not enough brood to replace the older bees, the adult worker population will sharply decline [46].

Fall temperatures that stimulate foraging in the fall can affect the workers themselves through the stress caused by flight. Bees physiologically age when they fly [48,49]. Extended periods of flight in the fall produce higher proportions of physiologically older short-lived workers in the winter cluster. When foraging resumes in the spring, loss of older workers can exceed numbers of newly emerged bees and the colony population can decline. The combination of limited brood rearing while bees are still confined in the hive due to low levels of food and high numbers of short-lived bees can set up a negative feedback loop when foraging resumes. Fewer bees are available to forage and to rear brood, so there are fewer bees to replace short-lived workers that overwintered. he population momentum spirals downward and can cause colony loss in the spring [50].

Adding to the challenges of overwintering colonies in years with warm fall is the presence of Varroa mites. This external parasite of honey bees can spread between colonies on foragers that drift among hives. Foragers entering colonies with mites on their bodies increase in the fall and can sharply increase Varroa populations [51]. The longer colonies can forage in the fall, the longer mites can migrate among hives on foragers and infest colonies. Colonies with high mite populations have little chance of surviving over the winter [52] because Varroa feed on the fat body of bees [53] and might compromise this organ that plays such prominent and diverse roles in overwintering. Furthermore, Varroa transmit viruses that can compromise the immune system of bees (see ref. [54]) and reduce their lifespans, adding to the numbers of short-lived bees in the winter cluster. In the next chapter, I will tell you more about Varroa and how parasites co-op systems in their hosts and use them to their advantage.

4.4 Final thoughts

A honey bee colony is a dynamic and resilient organism. Individual bees have a range of behaviors and capabilities that when performed in cooperation with nestmates allow colonies flexibility in responses to changing conditions across landscapes and seasons. The ability of colonies to sense changes in environmental conditions and respond appropriately is a reason why honey bees have established on every continent except Antarctica and in a range of habitats and elevations and from tropical forests to deserts. The multitude of activities occurring in a colony at any moment can, from a systems perspective, be seen as a coordination among numerous subsystems that orchestrate graded responses to outside disturbances and restore colony equilibrium. However, systems have boundaries around the ranges where they can respond. Climate change is presenting conditions that can exceed response boundaries. Although stress in the system from disturbances such as nutritional stress or extreme temperatures may be alleviated in the short term, there can be long-term consequences on colony growth, food storage, and overwintering success.

If organisms face conditions that negatively impact their survival and reproduction, they migrate to more suitable habitats. Migration of feral honey bees can occur from swarming. Habitats where swarms are likely to establish colonies that can survive conditions throughout the year and generate swarms of their own will likely shift geographically. Based on current predictions of suitable habitats for flowering plants and pollinators, those sites might become more limited [55]. There is a critical need for long-term conservation efforts and policies directed at safeguarding areas to support populations of honey bees and other pollinators as they move from current areas into more suitable landscapes.

Long-established apiary sites where high colony losses are now occurring with greater frequency might be transitioning to areas where conditions often are beyond the boundaries where colonies can adjust in ways that ensure growth and long-term survival. In the case of managed honey bee colonies, migration may need to take the form of beekeepers moving from current apiary sites that no longer support colony growth and survival. However, identifying new areas for apiaries can be difficult. The sites must be accessible, provide flowering plants from spring to fall, and have environmental conditions within the boundaries where colonies can respond and restore long-term growth and survival. Tools to aid beekeepers choose new apiary sites where colonies can thrive need to be developed and made available. These include landscape models of flowering plants and pollinator species distributions across space and time. The landscape models need to include expected climatic and environmental conditions to predict current and future suitability of the sites [56]. Information from land suitability assessments [55] also would be helpful to beekeepers. The assessments need to include information on pollen and nectar availability along with precipitation and temperature predictions to project potential frequencies of heat waves, droughts, and warmer fall and winter temperatures, especially in temperate areas. The assessments also could be useful in identifying areas where land is managed specifically for establishing pollinator populations.

References

[1] Winston ML. The biology of the honey bee. Cambridge, MA: Harvard University Press; 1987.

[2] Schulz DJ, Huang ZY, Robinson GE. Effects of colony food shortage on behavioral development in honey bees. Behav Ecol Sociobiol 1998;42(5):295–303.

[3] Toth AL, Kantarovich S, Meisel AF, Robinson GE. Nutritional status influences socially regulated foraging ontogeny in honey bees. J Exp Biol 2005;208(24):4641–9.

[4] Prado A, Requier F, Crauser D, Le Conte Y, Bretagnolle V, Alaux C. Honeybee lifespan: the critical role of pre-foraging stage. R Soc open Sci 2020;7(11):200998.

[5] DeGrandi-Hoffman G, Roth SA, Loper GL, Erickson Jr EH. BEEPOP: a honeybee population dynamics simulation model. Ecol Model 1989;45(2):133–50.

[6] Perry CJ, Søvik E, Myerscough MR, Barron AB. Rapid behavioral maturation accelerates failure of stressed honey bee colonies. Proc Natl Acad Sci U S A 2015;112(11):3427–32.

[7] Haydak MH. Honey bee nutrition. Annu Rev Entomol 1970;15:143–56.

[8] Schmickl T, Crailsheim K. How honeybees (Apis mellifera L.) change their broodcare behaviour in response to non-foraging conditions and poor pollen conditions. Behav Ecol Sociobiol 2002;51(5):415–25.

[9] Brodschneider R, Crailsheim K. Nutrition and health in honey bees. Apidologie 2010;41(3):278–94.

[10] Schmickl T, Crailsheim K. Cannibalism and early capping: strategy of honeybee colonies in times of experimental pollen shortages. J Comp Physiol A 2001;187(7):541–7.

[11] Schmickl T, Blaschon B, Gurmann B, Crailsheim K. Collective and individual nursing investment in the queen and in young and old honeybee larvae during foraging and non-foraging periods. Insects Soc 2003;50(2):174–84.

[12] Carroll MJ, Brown N, Huang E. E-B-ocimene and brood cannibalism: interplay between a honey bee larval pheromone and brood regulation in summer dearth colonies. PLoS ONE 2025;20(2):e0317668.

[13] He XJ, Zhang XC, Jiang WJ, Barron AB, Zhang JH, Zeng ZJ. Starving honey bee (Apis mellifera) larvae signal pheromonally to worker bees. Sci Rep 2016;6(1):22359.

[14] Webster TC, Peng Y-S. Passage of cannibalized tissue among honey bee (Hymenoptera: Apidae) colony members. Ann Entomol Soc Am 1987;80(6):814–9.

[15] Webster TC, Peng Y-S, Duffey SS. Conservation of nutrients in larval tissue by cannibalizing honey bees. Physiol Entomol 1987;12(2):225–31.

[16] Schmickl T, Crailsheim K. Inner nest homeostasis in a changing environment with special emphasis on honey bee brood nursing and pollen supply. Apidologie 2004;35(3):249–63.

[17] Palmer KA, Oldroyd BP. Evolution of multiple mating in the genus Apis. Apidologie 2000;31(2):235–48.

[18] Kronenberg F, Heller HC. Colonial thermoregulation in honey bees (Apis mellifera). J Comp Physiol 1982;148(1):65–76.

[19] Groh C, Tautz J, Rössler W. Synaptic organization in the adult honey bee brain is influenced by brood-temperature control during pupal development. Proc Natl Acad Sci U S A 2004;101(12):4268–73.

[20] Zhao H, Li G, Guo D, Li H, Liu Q, Xu B, et al. Response mechanisms to heat stress in bees. Apidologie 2021;52(2):388–99.

[21] Tautz J, Maier S, Groh C, Rossler W, Brockmann A. Behavioral performance in adult honey bees is influenced by the temperature experienced during their pupal development. Proc Natl Acad Sci U S A 2003;100(12):7343–7.

[22] Jones JC, Helliwell P, Beekman M, Maleszka R, Oldroyd BP. The effects of rearing temperature on developmental stability and learning and memory in the honey bee, Apis mellifera. J Comp Physiol A 2005;191:1121–9.

[23] Kablau A, Berg S, Härtel S, Scheiner R. Hyperthermia treatment can kill immature and adult varroa destructor mites without reducing drone fertility. Apidologie 2020;51:307–15.

[24] Simpson J. Nest climate regulation in honey bee colonies: honey bees control their domestic environment by methods based on their habit of clustering together. Science 1961;133(3461):1327–33.

[25] Southwick EE, Heldmaier G. Temperature control in honey bee colonies. BioScience 1987;37(6):395–9.

[26] Southwick EE, Moritz RF. Social control of air ventilation in colonies of honey bees, Apis mellifera. J Insect Physiol 1987;33(9):623–6.

[27] Bonoan RE, Iglesias Feliciano PM, Chang J, Starks PT. Social benefits require a community: the influence of colony size on behavioral immunity in honey bees. Apidologie 2020;51:701–9.

[28] Starks PT, Gilley DC. Heat shielding: a novel method of colonial thermoregulation in honey bees. Naturwissenschaften 1999;86(9):438–40.

[29] Jones JC, Myerscough MR, Graham S, Oldroyd BP. Honey bee nest thermoregulation: diversity promotes stability. Science 2004;305(5682):402–4.

[30] McAfee A, Chapman A, Higo H, Underwood R, Milone J, Foster LJ, et al. Vulnerability of honey bee queens to heat-induced loss of fertility. Nat Sustainability 2020;3:367–76.

[31] Mitchell D. Honeybee cluster—not insulation but stressful heat sink. J R Soc Interface 2023;20:20230488.

[32] Stabentheiner A, Pressl H, Papst T, Hrassnigg N, Crailsheim K. Endothermic heat production in honeybee winter clusters. J Exp Biol 2003;206(2):253–8.

[33] Mattila HR, Harris JL, Otis GW. Timing of production of winter bees in honey bee (Apis mellifera) colonies. Insectes Sociaux 2001;48(2):88–93.

[34] Kunc M, Dobeš P, Hurychová J, Vojtek L, Beani Poiani S, Danihlik J, et al. The year of the honey bee (Apis mellifera L.) with respect to its physiology and immunity: A search for biochemical markers of longevity. Insects 2019;10(8).

[35] Kunert K, Crailsheim K. Seasonal changes in carbohydrate, lipid and protein content in emerging worker honeybees and their mortality. J Apic Res 1988;27(1):13–21.

[36] Amdam GV, Norberg K, Hagen A, Omholt SW. Social exploitation of vitellogenin. Proc Natl Acad Sci U S A 2003;100(4):1799–802.

[37] Döke MA, Frazier M, Grozinger CM. Overwintering honey bees: biology and management. Curr Opin Insect Sci 2015;10:185–93.

[38] Münch D, Amdam GV. The curious case of aging plasticity in honey bees. FEBS Lett 2010;584:2496–503.

[39] Amdam GV. Social context, stress, and plasticity of aging. Aging Cell 2011;10:18–27.

[40] Lee S, Kalcic F, Duarte IF, Titera D, Kamler M, Mrna P, et al. 1H NMR profiling of honey bee bodies revealed metabolic differences between summer and winter bees. Insects 2022;13(2):193.

[41] Arrese EL, Soulages JL. Insect fat body: energy, metabolism, and regulation. Annu Rev Entomol 2010;55:207–25.

[42] Teulier L, Weber JM, Crevier J, Darveau CA. Proline as a fuel for insect flight: enhancing carbohydrate oxidation in hymenopterans. Proc R Soc B Biol Sci 2016;283(1834):20160333.

[43] Koubová J, Sábová M, Brejcha M, Kodrík D, Čapková Frydrychová R. Seasonality in telomerase activity in relation to cell size, DNA replication, and nutrients in the fat body of Apis mellifera. Sci Rep 2021;11(1):1–12.
[44] Aubert G, Lansdorp PM. Telomeres and aging. Physiol Rev 2008;88:557–89.
[45] DeGrandi-Hoffman G, Corby-Harris V, Graham H, Watkins deJong E, Chambers M, Snyder L. The survival and growth of honey bee (Hymenoptera: Apidae) colonies overwintered in cold storage: the effects of time and colony location. J Econ Entomol 2023;116(4):1078–90.
[46] Rajagopalan K, DeGrandi-Hoffman G, Pruett M, Jones VP, Corby-Harris V, Pireaud J, et al. Warmer autumns and winters could reduce honey bee overwintering survival with potential risks for pollination services. Sci Rep 2024;14(1450):5410.
[47] Kammerer M, Goslee SC, Douglas MR, Tooker JF, Grozinger CM. Wild bees as winners and losers: relative impacts of landscape composition, quality, and climate. Glob Change Biol 2021;27(6):1250–65.
[48] Williams JB, Roberts SP, Elekonich MM. Age and natural metabolically-intensive behavior affect oxidative stress and antioxidant mechanisms. Exp Gerontol 2008;43:538–49.
[49] Margotta JW, Mancinelli GE, Benito AA, Ammons A, Roberts SP, Elekonich MM. Effects of flight on gene expression and aging in the honey bee brain and flight muscle. Insects 2013;4(1):9–30.
[50] Betti MI, Wahl LM, Zamir M. Age structure is critical to the population dynamics and survival of honeybee colonies. R Soc Open Sci 2016;3(11).
[51] DeGrandi-Hoffman G, Ahumada F, Zazueta V, Chambers M, Hidalgo G, Watkins deJong E. Population growth of Varroa destructor (Acari: Varroidae) in honey bee colonies is affected by the number of foragers with mites. Exp Appl Acarol 2016;69(1):21–34.
[52] Genersch E, Von der Ohe W, Kaatz H, Schroeder A, Otten C, Büchler R, et al. The German bee monitoring project: a long term study to understand periodically high winter losses of honey bee colonies. Apidologie 2010;41(3):332–52.
[53] Ramsey SD, Ochoa R, Bauchan G, Gulbronson C, Mowery JD, Cohen A, et al. Varroa destructor feeds primarily on honey bee fat body tissue and not hemolymph. Proc Natl Acad Sci U S A 2019;116(5):1792–801.
[54] DeGrandi-Hoffman G, Chen Y. Nutrition, immunity and viral infections in honey bees. Curr Opin Insect Sci 2015;10:170–6.
[55] Tennakoon S, Apan A, Maraseni T. Unravelling the impact of climate change on honey bees: an ensemble modelling approach to predict shifts in habitat suitability in Queensland, Australia. Ecol Evolut 2024;14:e11300.
[56] Guisan A, Rahbek C. SESAM—a new framework integrating macroecological and species distribution models for predicting spatio-temporal patterns of species assemblages. J Biogeogr 2011;38(8):1433–44.

Chapter 5

Colony as a host for parasites and pathogens: The fusion of Varroa and honey bee colony population dynamics

In Chapter 2, I briefly discussed how Varroa mites threaten honey bee colony growth and survival. In this chapter, I will go into depth about how this parasite causes colony losses. Varroa's impact on colonies has changed greatly since its shift from Asian honey bee (*Apis cerana*) into European honey bee (*Apis mellifera* L.) colonies. In the beginning, Varroa was controlled in *A. mellifera* by well-timed miticide treatments. The picture has changed, however, due to colony management and mite control methods, virus transmission by Varroa, and climate change. These factors have changed components in the system Varroa employs to survive and disperse among colonies and have made the mite more virulent.

Before I tell the story of Varroa, I want to present a broader picture of parasites. These organisms are masters at building systems with strategies that set them up for long-term success. Common strategies employed by parasites are manipulating host behaviors, parasitizing the most vulnerable life stages of their hosts, and evolving ways to spread to new hosts (i.e., dispersal mechanisms). Varroa employs all these strategies, but a difference between Varroa and many other parasites is that Varroa's host is the colony and not just the individual bee. Because the colony is the host, Varroa's population dynamics are intimately tied to the annual cycle of the colony. How the cycle is shaped by management methods and climate change drives the dynamics of Varroa populations and creates conditions for either Varroa coexistence with the bees or colony failure and mite dispersal to new hives.

5.1 Overview of parasites

Parasites are everywhere and compose nearly half of animal life on Earth (see refs [1,2]). Parasitism is a form of symbiosis; that is, two different organisms living in close physical association usually to the advantage of both. Symbiotic relationships are found between and among all life forms. Honey bees and

The Honey Bee and Climate Change. DOI: https://doi.org/10.1016/B978-0-443-28995-8.00007-6

other organisms have symbiotic relationships with various bacteria and fungi (i.e., microbes) that benefit both groups. The microbes aid in processes such as food digestion and preservation. The microbes benefit because bees provide living space within their bodies and the hive, food sources, and means of transport [3]. Flowering plants and their pollinators also have a symbiotic relationship. Flowers provide food for pollinators (nectar and pollen) and pollinators collect it, and in the process, transfer pollen and initiate seed set.

Parasitic relationships differ from symbiotic ones in that one organism benefits at the expense of another. Parasites use other organisms (the host) as both living space and food sources. In some instances, the host and parasite coexist. This relationship can occur when the parasite needs the host to successfully reproduce so that it can move from the parent (original host) to their offspring (new host). In these cases, the parasite inflicts minimum damage to the host. The host coexists with the parasite and provides a steady source of new hosts for their offspring. There are cases, however, where parasites harms or even kills their host. These parasites must develop means to leave their host and find another to infect (dispersal mechanisms). When a parasite can successfully disperse and infect another host, highly virulent strains of the parasite can evolve.

Parasites have developed systems and strategies that increase their chances of successfully invading a host. One strategy employed by many species of parasites is to target vulnerable life stages or conditions in the host so the parasite can circumvent immune responses that might thwart their invasion. Young individuals are particularly good targets for parasites as the high energetic demands of growth can reduce immune responses [4]. Parasites also can exploit brood care behaviors to their advantage [5]. In these cases, individuals can acquire parasites during development and are usually more severely affected than if they were parasitized as adults.

Parasites that harm their hosts and ultimately cause their death must develop subsystems for dispersal. Those dispersal subsystems can be plugged into the host's behavior to increase the frequency of encountering with new hosts. A classic case of host manipulation to aid and abet dispersal involves a parasitic trematode (*Leucochloridium* spp.), a snail (specifically "lighthouse snails" (*Succinea* spp.)), and birds. Trematodes are flat leaf-like organisms commonly called flat worms or flukes. The cycle starts when lighthouse snails eat bird droppings containing eggs of the parasitic trematode. The eggs hatch in the snail. As the trematode grows, the color and shape of the snail's tentacles change and begin pulsating in response to light. The color and pulsating light look like an insect larva and attracts birds. The birds consume the snail and become infected by the trematode. The trematode establishes in the digestive tract of the bird and produces eggs that are released into the bird's feces. Snails eat the feces, and begin the cycle again [6]. The lighthouse snail parasite, and many other parasites, affect host behaviors by altering host physiology rather than producing compounds on their own to trigger the behaviors that enhance its dispersal.

As we move on to Varroa, I will describe how the mite parasitizes vulnerable life stages, exploits brood care behaviors, proliferates as the colony (i.e., host), grows, and disperses by changing honey bee worker behavior. You also will see that virulence and mite dispersal can be the product of environmental conditions. In cases where colonies are widely separated as in undisturbed ecosystems such as forests, the mite is most likely to be transferred vertically (i.e., among the bees in the colonies). Damage from the mite is low because host reproduction (swarming) is needed for the parasite to survive and disperse. However, in cases where Varroa can transfer horizontally (i.e., among colonies), highly virulent mite populations can develop, leading to colony collapse, mite dispersal, and establishment of mites in new colonies.

5.1.1 Varroa mites

Before going into the population dynamics of honey bees and Varroa in a colony, I need to provide a general overview of how Varroa operates. Varroa mites are ectoparasites (i.e., they live on the outside of the host's body) of worker and drone honey bees. They parasitize adults, larvae, and pupae [7] (Fig. 5.1). Varroa originated in Asia where it parasitizes the Asian honey bee, *Apis cerana* Fabr. Varroa levels remain low in *A. cerana* colonies because the mite reproduces only in drone brood. This limits host numbers, since drone brood comprises <5% of the brood population. *A. cerana* workers also remove and kill mites attempting to parasitize adult bees and brood ("grooming behavior") [8,9].

FIGURE 5.1 Varroa mites on the body of a pupa and adult worker. Varroa mite (*Varroa destructor*) parasitizing a worker pupa (*left*) an adult worker bee (*right*).
Photographs by Mona Chambers (left) *and Emily Watkins deJong* (right).

In the 1950s, Varroa expanded its host range to include the European honey bee (*Apis mellifera* L.). Varroa was found in *A. mellifera* colonies in eastern Russia in 1952 and then Japan in 1957 (see ref. [10]). From there, the mite spread to Europe and was found in colonies in Bulgaria in 1967, Germany (1971), and Romania (1975). Varroa spread throughout Europe and was first detected in the United Kingdom in 1992. Although regulatory measures were enacted to prevent Varroa from spreading to the United States, in 1987, Varroa were detected in Florida, and throughout the United States by the 1990s. Varroa were found in South Africa in 1997, New Zealand in 2000, and Australia in 2022. The expansion of Varroa into *A. mellifera* colonies has been successful because the mite can parasitize both workers and drones, enabling the parasite to have multiple hosts in the colony. In addition, the ability to parasitize worker brood increased the numbers of possible hosts in the colony. Colonies rear far more worker than drone brood over longer periods during the year, so mite population growth also increased. A second factor contributing to Varroa's success is that grooming behavior is not as prevalent or efficient in *A. mellifera* as in *A. cerana*, so removing mites while feeding or reproducing is less likely.

5.1.2 Varroa reproduction

As mentioned earlier, parasites increase their success by attacking hosts during vulnerable life stages particularly immatures. Parasites also can exploit brood care to increase their chances of finding a suitable host and parasitizing it with minimal risk. Varroa implements both of these strategies. First, Varroa parasitizes worker and drone larvae and pupae. The female mite (called mother mite or foundress) enters brood cells just before they are sealed [7]. The mother mite starts to feed on the nutrient rich fat body of the developing bee larva within 6 h of the cell being sealed [11,12]. The wound in the larva created by the mother mite becomes the site for her continued feeding and that of her offspring.

Offspring are produced by mother mites shortly after she begins to feed on the larva. The first egg the foundress lays develops into a male mite. The next egg develops into a female mite that mates with the male. In worker brood, foundress mites produce 1 to 1.5 mated daughter mites. In drone brood, 2–3 daughter mites can be produced because drones take longer than workers to mature into adults. When adult workers and drones emerge from their cells, the foundress mite and her daughters emerge too. The male dies in the cell while the mated daughter mites can enter new brood cells as foundresses and start the cycle again [7].

The second general strategy parasites use to be successful is exploiting brood care. In the case of Varroa, mother and daughter mites that emerge from brood cells attach to young worker bees that are tending developing brood (i.e., nurse bees). Mites can discriminate between young nurse bees and older workers by the differences in chemical compounds on the bee's outer surface (i.e., cuticle) [13,14]. Mites carried on adult bees, are in a "phoretic" stage (derived from the Greek verb "phero" meaning 'to carry'). When the mite is carried to a brood cell of suitable age, the mite detaches and enters the cell to reproduce.

Geographic location strongly influences the growth of Varroa populations throughout the year and their impact on colonies. In southern latitudes and areas with desert or tropical climates, brood rearing can occur year around, so there is no break in Varroa reproduction. In temperate climates, brood rearing slows and eventually stops in the fall as the bees prepare for confinement in the hive over the winter. The break in brood rearing halts Varroa reproduction. However, Varroa can overwinter in a phoretic stage on bees in the winter cluster and feed on them while confined in the hive. When brood rearing resumes, the mites infest the first brood cells that are reared. This is significant as the brood are replacements for older bees in the winter cluster. For the colony to grow in the spring, the brood reared while the bees are still in cluster need to be healthy and live long enough to replace the older bees in the winter cluster. Rearing healthy long-lived bees is difficult when Varroa numbers are high because the mite removes nutrients from the fat body, compromises the bee's immune system, and transmits viruses.

5.1.3 Varroa and viruses

Parasitism by Varroa is stressful to bees, but colony losses, especially during the winter, are caused most often by the viruses Varroa transmit [15]. A foundress mite feeding on a developing larva transmits and activates several different single-stranded RNA viruses including acute bee paralysis virus, ABPV; black queen cell virus; Israeli acute paralysis virus; Kashmir bee virus, Sacbrood bee virus; and deformed wing virus, DWV (see ref. [16]). Bees also can spread viruses to mites as occurs when phoretic mites feed on adult bees. Those viruses then can be spread to other bees when the parasite feeds [17]. Throughout the spring and summer, bees may be infected with several viruses, but in most cases by the fall, DWV is the predominant virus detected [18].

DWV is ubiquitous in colonies and occurs in all life stages of honey bees. The virus is in eggs because DWV is carried in sperm and transmitted from the queen to fertilized eggs [19]. The transfer of DWV from sperm to fertilized eggs causes DWV to persist in colonies even if Varroa is absent. The transmission of DWV in sperm also makes drones agents of virus dispersal as they transmit the virus through their sperm to new colonies when mating with virgin queens.

There are many other ways that DWV can be spread throughout the colony. Honey, pollen, and royal jelly can carry DWV. Contact with feces from larvae containing DWV, cannibalism of developing larvae and pupae, and food exchange between bees (trophallaxis) also can spread DWV (see ref. [4]).

Levels of DWV can be comparatively low as the colony begins its yearly cycle in the spring but increase greatly by the fall especially if the colony is infested with Varroa [20]. High DWV titers in bees can cause fatal wing deformities (Fig. 5.2) and when they occur in the fall are a harbinger of colony loss over the winter [18].

The relationship between Varroa and viruses especially DWV has been mutually beneficial. Varroa parasitize immature stages of honey bees, and these are more susceptible to DWV infection than adults [21]. DWV reduces immune responses particularly those involved with wound healing [22]. Insects heal wounds to their cuticle by a process called melanization. The process involves secreting a compound called melanin to form a hard covering over the damaged cuticle. Reducing melanization in larvae infected with DWV benefits Varroa by preventing the larva from forming a hard cover over the puncture in the cuticle made by the mother mite. The open wound enables the

FIGURE 5.2 **Worker honey bees with normal wings and deformed wings due to deformed wing virus.** A worker honey bee that has normal, fully formed functional wings and a worker bee that was infected with deformed wing virus during the larval and pupal stages. The infected worker emerged from the cell with permanently deformed wings that will prevent her from flying. *Photographs by Mona Chambers.*

mother mite and her offspring to feed continuously on the developing bee and in the process, transmit virus while under the capped cell.

A final word about DWV. Although DWV is a single type of virus, there are at least four major variants (DWV-A, B, C, and D). DWV-B causes most of the symptoms commonly seen in adult bees infected with DWV. These are in addition to deformed wings, bloated abdomen, paralysis, and a severely reduced lifespan [23].

5.1.4 Controlling Varroa populations

Because of their lifestyle, Varroa is a very difficult parasite to control with miticides. At any time, most of the mite population can be under capped brood cells and out of reach of most miticides. For this reason, the target for most miticides is the Varroa on the bodies of adult bees. However, this is a moving target. The mites on adult bees are moving into cells whenever bees are rearing brood. Mites also are emerging from cells with the bees they parasitized. For all the mites in a colony to be exposed to a miticide, the active ingredient must remain in the colony at levels that are lethal to mites for at least 2–3 weeks. This period of effectiveness will ensure that all mites on adult bees and those emerging from capped worker and drone brood have a chance of being exposed to the miticide.

There are several miticides available to control Varroa. Miticides differ in formulation and delivery. Formic acid, thymol, and oxalic acid work as vapors and kill phoretic mites [7]. Formic acid is unique among the vapor-based miticides, as it also can kill mites parasitizing pupae in sealed brood cells [24]. The effectiveness of vapor-based miticides and whether they can harm the bees, brood, and queen depends on ambient temperatures [25,26]. If temperatures are high such as during heat waves, miticides that are delivered as vapors can cause queen loss and injure the brood.

There also are miticides that are applied in strips that bees walk on and acquire by contact. As bees interact with nestmates and make body-to-body contact, the miticide can be spread throughout the colony in much the same way as pollen and contact pheromones are distributed.

The first miticides used against Varroa (tau-fluvalinate and coumaphos) were contact miticides. Because these miticides were used repeatedly to control Varroa, over time the mite became resistant [27]. The latest contact miticide to become available is amitraz. However, resistance to amitraz now is occurring [28], but might be colony specific as the effectiveness of amitraz seems to vary among colonies. Resistant mites could spread among hives from mite migration (see below) and establish new mite-resistant populations in colonies where mites were once susceptible.

The impact that resistance to amitraz might have on managed colonies was evident in the 2024–2025 season. During that time, managed colony losses exceeded 60%. DWV and ABPV were found in bees from the dying colonies.

Both viruses are transmitted by Varroa, and, interestingly, the Varroa in all the dying colonies were resistant to amitraz (https://www.biorxiv.org/content/10.1101/2025.05.28.656706v1).

Another means of managing Varroa is mite-resistant or tolerant lines of bees that keep mites at low levels throughout the year and reduce the need for miticides. Resistant lines keep Varroa levels low because the bees remove mites from infested cells (hygienic behavior) and from the bodies of nestmates (grooming behavior) [29]. Other genetic lines suppress mite reproduction [30,31]. While Varroa levels can be kept low during most of the year, colonies with resistant bees might still fall victim to mite migration so that by late fall, mite numbers are comparable to unselected lines [32]. Mite-resistant lines could be improved if there also was selection for nestmate recognition that prevented foragers from other colonies from entering the hive.

5.2 Mite migration

When Varroa was first discovered in managed colonies, it took 3–4 years for mite populations to grow to levels that threatened the colony's survival [33]. In recent years, however, colonies with low mite populations in the spring can die from heavy mite loads by fall or over the winter. Rapid increases in mite numbers over short periods of time cannot be explained by mite reproduction, which is relatively low especially in worker brood where most mites are found [34]. At best, 1–2 mated daughter mites are reared per mother mite parasitizing worker brood, and she successfully reproduces in only about 80% of the cells she infests [7]. Although more mated mites are produced in drone cells, drones comprise, at best, about 5% of the colony. Further limiting mite population growth is an inherent reduction in reproduction when Varroa outnumber cells to infest. By this I mean that when there are more mother mites than suitable brood cells, multiple mites will enter each cell. Reproduction and development of new daughter mites is reduced with each additional mother mite that enters a brood cell. This is especially important in drone cells that can be multiply infested because they are highly attractive to mother mites. The result is that mite populations grow more slowly when Varroa populations are high and larvae of suitable age are limited (see ref. [7]). So, what explains the sharp increases in mite populations in colonies? A tenet of population ecology is that if populations grow at rates that exceed projections from reproduction, individuals might be moving in from the outside. In other words, there is migration.

There is both circumstantial and direct evidence for Varroa migration into colonies. For instance, colonies without mites established in isolated areas eventually become infested with Varroa [35]. There also is evidence of foragers with mites on their bodies entering colonies that were not their own [36] (Fig. 5.3). As foragers with mites entering colonies increases, so does the Varroa population in the colony [34].

FIGURE 5.3 **Honey bee pollen forager carrying Varroa mites.** Pollen foraging honey bee captured while entering a honey bee hive. The forager is carrying two Varroa mites on the ventral surface of her abdomen. Foragers carrying mites and drifting to hives that are not their own might account for sharp increases in Varroa populations that cannot be accounted for by mite reproduction alone.
Photograph by Mona Chambers.

Although some foragers carrying Varroa might be from colonies in the same or nearby apiaries, the foragers also can come from colonies outside the apiary. Mite-free colonies that were 1–1.5 km (0.6–1 miles) away from the nearest hives became infested with mites [35,36]. The source of mites was thought to be bees emigrating from highly infested colonies referred to as 'mite bombs' or from colonies that were collapsing from Varroa infestations. Varroa would attach to bees robbing the collapsing colonies and be transported back to the robber's hive.

There is little evidence to support the mite bomb theory, though colonies collapsing from mites and being robbed does seem to be a plausible source of mites entering colonies [35,37]. Foragers that find a colony to rob could cause a sharp increase in mites over a short period of time. However, this is not always the case. Studies measuring the frequency of collecting foragers with mites entering colonies indicate that they comprise a small percentage of all foragers entering colonies [20,34]. More often the increase in Varroa might be from an accumulation of mites trickling into the colony over time. Much like putting a bucket under a leaky faucet, it may seem like a small amount of water is dripping but over time, the water accumulates and eventually can overflow the bucket.

Varroa attaching to foragers and moving to other colonies is a form of dispersal. However, several coordinated behaviors need to occur for the mite to disperse successfully. First, mites have to attach to foragers instead of nurse bees. Remember, mites use chemical cues on the bee's cuticle to discriminate between young nurse bees and older workers [13,14]. Varroa prefer attaching to nurse bees because they can transport the mite to brood cells where reproduction occurs.

A second requirement for successful dispersal is that foragers with mites have to drift to another colony when returning from a foraging flight. Precocious foragers (i.e., worker bees that forage before they are 2–3 weeks old) could be vehicle to transport mites to other colonies. If the worker was young when she began to forage, she may still have chemical cues so that mites attach to her. Precocious foragers also have a high frequency of drifting to other hives due to incomplete cognitive development [38].

The combination of Varroa and viruses they transmit, especially DWV, might play a role in the migration/dispersal of Varroa. Instances where viruses alter insect behaviors to increase their dispersal have been documented in many species. A well-known case in that of Spongy moth (*Lymantria dispar* (L.)) larvae infected with the baculovirus, Lymantria dispar nucleopolyhedro virus. Spongy moth larvae infected with the virus climb to the top of the trees where they are feeding and become black liquid mass due to the virus infection. As the larvae liquifies on exposed upper limbs, millions of infective virus particles are released into the air spreading the virus throughout the forest [39].

There are reasons to suspect that viruses could be altering behaviors of adult worker bees in ways that aid Varroa and virus dispersal. Larvae that are parasitized by Varroa and infected with viruses, especially DWV, begin foraging earlier in adult life [40]. Precocious foragers and those infected with DWV are more likely to drift to other hives when returning from a foraging flight. This is because in addition to incomplete neural development in young bees, DWV infections impair worker bee's memory and cognitive ability [41,42]. The combination of young workers that can attract mother mites in the hive and virus-infected precocious foragers having poor navigation skills could create a scenario where workers from mite-infested colonies could spread Varroa among hives throughout their foraging area.

Varroa migration into new colonies can happen at any time of year when bees are flying, but it seems to increase in the fall when virus titers are at their highest levels [34,43]. Dispersal in the fall would be advantageous to the mite because colonies with high Varroa and virus loads often die over the winter

[44]. If the colony dies, mites and viruses die, too. Mite dispersal in the fall could save Varroa and the viruses they transmit from perishing over the winter when temperatures prevent bees from flying and mites from dispersing.

Although the movement of mites among colonies is well documented, we do not know if foragers use cues to find the hives they enter. For mite dispersal to be successful, the hive the forager drifts into must have a better chance of surviving than the forager's original colony. Since foragers carrying mites can drift beyond the apiary where they originated, it may be that odors emanating from hives serve as cues to drifting foragers. In our study measuring mite migration into colonies, we found that some hives have more incoming foragers with mites than others, suggesting that foragers with mites may have been keying in on them [34]. Controlling immigration of mites into colonies is essential for keeping mite numbers low. For this to be possible, however, we need to understand fully the mite dispersal system that causes Varroa to attach to foragers and for them to enter new hives.

5.2.1 Honey bee–Varroa relationship in feral and managed colonies

The risk Varroa poses to colony survival can be quite different between feral and managed colonies. Managed colonies can be lost to Varroa in less than a year, particularly if mite numbers are high in the fall. In contrast, feral colonies can persist for years and maintain low levels of Varroa [45,46]. Although no resistance mechanisms have been uncovered to explain how Varroa numbers are kept low in feral colonies, two factors might be responsible: colony distribution across the landscape and whether the colony can swarm [43]. These factors differ greatly between managed and feral colonies and might set the stage for colonies either coexisting or succumbing to Varroa.

Let us begin with the importance of colony distribution. Apiaries can contain dozens if not hundreds of hives separated by only a few meters. Feral colonies often exist at low densities (1 colony per km^2 (about 0.4 $miles^2$)) and reside in well-concealed cavities of trees or rock crevasses. In forest and mountain landscapes, the nearest-neighbor distance between feral hives averages about 850 m (about 0.5 miles) [45,47]. If Varroa causes a managed colony to collapse, nearby colonies in the apiary could provide new hosts. There is a good chance the collapsing colony will be discovered and robbed by foragers from other colonies. The mites could attach to the robbers and be taken to new hives. In the case of feral colonies, however, the chances of discovery by a nearby colony probably is much lower than in an apiary. If Varroa causes a feral colony to collapse, the mites undoubtedly will perish too. The difference in colony distribution would put different selective pressures on the mite where traits that could collapse a colony would be selected against in feral colonies, but might persist in managed ones. Currently, in apiaries, Varroa is very difficult to control, supporting a general attribute of parasites that when they can successfully disperse and infect another host, highly virulent strains can evolve.

The second factor that differs between feral and managed colonies is the frequency of swarming. When a feral colony outgrows the cavity where it resides, the colony swarms. In managed colonies, swarming and the loss of thousands of bees are prevented because it slows colony growth and reduces honey production. When colonies become crowded and bees are running out of space for brood and food stores, beekeepers add frames of empty comb to relieve overcrowding.

So why would swarming affect Varroa population growth? Varroa reproduce in brood cells. Colonies that are preparing to swarm stop rearing brood, and this halts mite reproduction. The mite population is further reduced because some workers (15%–20%) that leave with the swarm carry phoretic mites [47]. In this way, swarming can be considered a means of dispersal for Varroa and a mite control method for the colony. The dispersal strategy, however, relies on Varroa having negligible effects on colony health, since only large vigorous colonies swarm. The benefits of annual swarming have been documented in feral colonies. These colonies that annually swarm have fewer Varroa, less disease, and higher survival rates than colonies that do not swarm [45,48].

5.2.2 Climate change and Varroa migration

Fall temperatures have increased by an average 2.7°F across the contiguous United States since 1970 (https://www.noaa.gov/news/planet-just-had-its-warmest-october-on-record). The warmer temperatures have extended summer-like conditions into the fall and delayed the cold needed for dormancy and hibernation. So, how might warmer temperatures affect the severity of Varroa infestations in colonies? One way that warmer fall temperatures may increase the damage caused by Varroa is that warmer conditions extend the period of brood rearing and as a consequence, the time when mites can reproduce [49]. Warmer fall temperatures also extend the period when bees will fly. Whenever bees are foraging, mites can be migrating among colonies. Although migration can occur throughout the spring and summer, foragers with mites are seen most often in the fall [34].

Migration can cause sharp increases in mite numbers at any time of year, but the consequences for colony survival might be the most serious in the fall when colonies are preparing for overwintering. The mites entering colonies and parasitizing the adult bees and brood are those that will comprise the winter cluster. Parasitism and possible virus transmission reduces the longevity of winter bees. Colonies with short-lived bees in the winter cluster are likely to die in the spring when foraging resumes [50].

Mites also can reduce the available nutrients in the fat body of overwintering beea, since this organ is the feeding site for the mite [11]. Nutrients in the fat body are the source of subsistence for the overwintering bees and are used to rear brood before foraging resumes in the spring. Therefore, high numbers of Varroa in the fall undermine the two essential components for successful overwintering; long-lived bees in the winter cluster and vigorous brood rearing while colonies are still confined in the hive. If overwintering bees are short-lived, and brood rearing is limited because of parasitism, bees would be dying without adequate numbers of replacements. In this case, the negative feedback loop between brood rearing and adult bees would be established that could cause colony loss in the spring.

Warmer falls have made it more difficult to control Varroa and manage colonies so that they will survive the winter. Preparations for overwintering in temperate areas need to be reimagined with consideration of the possibility of extended periods of flight weather in the fall. Monitoring for mites as long as flight weather occurs is a necessity because Varroa populations can increase rapidly. Solutions such as putting colonies in cold storage in the fall for overwintering might alleviate some losses, as removing hives from the environment stops foraging and mite migration. While this solution could be effective in areas with a temperate environment, it likely will not work for all beekeepers. For example, colonies in southern latitudes where brood rearing continues through the fall and winter do poorly when put into cold storage perhaps because bees have not transitioned from summer to winter phenotypes [51]. In these areas, colonies will need frequent monitoring for mites and treatment with miticides to keep Varroa populations low.

Varroa and the viruses they carry are parasites, and parasites are very good at adapting. When conditions change, parasites alter their strategies and find new ways to survive. Sometimes that involves changing behaviors or co-opting other organisms to provide assets the parasite alone does not possess. The adaptations Varroa has acquired in managed colonies and apiaries make it difficult to control using the same tactics of well-timed miticide treatments that were effective when Varroa first entered managed colonies. Colonies now have increases in both mite numbers and viral infections that lead to dispersal and Varroa becoming a migratory pest. Climate change has exacerbated the situation by prolonging periods of mite reproduction and dispersal. Colony losses from Varroa can be reduced but only if mite control strategies are adapted to address current climate conditions and the type of pest Varroa has become.

References

[1] Weinstein SB, Kuris AM. Independent origins of parasitism in Animalia. Biol Lett 2016;12:20160324.

[2] Poulin R. The many roads to parasitism: a tale of convergence. Adv Parasitol 2011;74:1–40.

[3] Steffan SA, Dharampal PS, Kueneman JG, Keller A, Argueta-Guzmán MP, McFrederick QS, et al. Microbes, the 'silent third partners' of bee–angiosperm mutualisms. Trends Ecol Evolut 2024;39(1):65–77.

[4] Zanni V, Frizzera D, Marroni F, Seffin E, Annoscia D, Nazzi F. Age-related response to mite parasitization and viral infection in the honey bee suggests a trade-off between growth and immunity. PLoS ONE 2023;18(7):e0288821.

[5] Rabeling C. Social parasitism. Encyclopedia of social insects. Springer International Publishing; 2020. p. 836–58.

[6] Heil M. Host manipulation by parasites: cases, patterns, and remaining doubts. Front Ecol Evolut 2016;4:80–182771.

[7] Rosenkranz P, Aumeier P, Ziegelmann B. Biology and control of Varroa destructor. J Invertebrate Pathol 2010;103:96–119.

[8] Fuchs S. Preference for drone brood cells by Varroa jacobsoni Oud in colonies of Apis mellifera carnica. Apidologie 1990;21:193–9.

[9] Peng Y-S, Fang Y, Xu S, Ge L. The resistance mechanism of the Asian honey bee, Apis cerana Fabr., to an ectoparasitic mite, Varroa jacobsoni Oudemans. J Invertebr Pathol 1987;49:54–60.

[10] Reams T, Rangel J. Understanding the enemy: a review of the genetics, behavior and chemical ecology of Varroa destructor, the parasitic mite of Apis mellifera. J Insect Sci 2022;22(1):1–10.

[11] Ramsey SD, Ochoa R, Bauchan G, Gulbronson C, Mowery JD, Cohen A, et al. Varroa destructor feeds primarily on honey bee fat body tissue and not hemolymph. Proc Natl Acad Sci U S A 2019;116(5):1792–801.

[12] Han B, Wu J, Wei Q, Liu F, Cui L, Rueppell O, et al. Life-history stage determines the diet of ectoparasitic mites on their honey bee hosts. Nat Commun 2024;15:725.

[13] Del Piccolo F, Nazzi F, Della Vedova G, Milani N. Selection of Apis mellifera workers by the parasitic mite Varroa destructor using host cuticular hydrocarbons. Parasitology 2010;137:967–73.

[14] Cervo R, Bruschini C, Cappa F, Meconcelli S, Pieraccini G, Pradella D, et al. High Varroa mite abundance influences chemical profiles of worker bees and mite-host preferences. J Exp Biol 2014;217:2998–3001.

[15] Schüler V, L Y-C, Gisder S, Horchler L, Groth D, Genersch E. Significant, but not biologically relevant: Nosema ceranae infections and winter losses in honey bee colonies. Commun Biol 2023;6(1):229.

[16] Grozinger CM, Flenniken ML. Bee viruses: ecology, pathogenicity, and impacts. Annu Rev Entomol 2019;64:205–26.

[17] Lamas ZS, S S, Ryabov EV, Mowery J, Heermann M, Sonenshine D, et al. Promiscuous feeding on multiple adult honey bee hosts amplifies the vectorial capacity of Varroa destructor. PLoS Pathog 2023;19(1):e1011061.

[18] Mondet F, d.M. JR, Kretzschmar A, Le Conte Y, Mercer AR. On the front line: quantitative virus dynamics in honeybee (Apis mellifera L.) colonies along a new expansion front of the parasite Varroa destructor. PLoS Pathog 2014;10(8):e1004323.

[19] Amiri E, Meixner MD, Kryger P. Deformed wing virus can be transmitted during natural mating in honey bees and infect the queens. Sci Rep 2016;6(1):33065.

[20] DeGrandi-Hoffman G, Corby-Harris V, Chen Y, Graham H, Chambers M, Watkins deJong E, et al. Can supplementary pollen feeding reduce varroa mite and virus levels and improve honey bee colony survival? Exp Appl Acarol 2020;82(4):455–73.

[21] J.R. de Miranda, E.G., Deformed wing virus. J Invert Path, 103, 2010, S48-S61

[22] Zhao Y, Heerman M, Peng W, Evans JD, Rose R, DeGrandi-Hoffman G, et al. The dynamics of deformed wing virus concentration and host defensive gene expression after Varroa mite parasitism in honey bees, Apis mellifera. Insects 2019;10(1):E16.
[23] Martin SJ, Brettell LE. Deformed wing virus in honeybees and other insects. Annu Rev Virol 2019;6:12.1–12.21.
[24] VanEngelsdorp D, Underwood RM, Cox-Foster DL. Short-term fumigation of honey bee (Hymenoptera: Apidae) colonies with formic and acetic acids for the control of Varroa destructor (Acari: Varroidae). J Econ Entomol 2008;10(2):256–64.
[25] Steube X, Beinert P, Kirchner WH. Efficacy and temperature dependence of 60% and 85% formic acid treatment against Varroa destructor. Apidologie 2021;52(3):720–9.
[26] Giacomelli A, Pietropaoli M, Carvelli A, Iacoponi F, Formato G. Combination of thymol treatment (Apiguard®) and caging the queen technique to fight Varroa destructor. Apidologie 2016;47(4):606–16.
[27] Mitton GA, Meroi Arcerito F, Cooley H, Fernandez de Landa G, Eguaras MJ, Ruffinengo SR, et al. More than sixty years living with Varroa destructor: a review of acaricide resistance. Int J Pest Manag 2022:1–18.
[28] Rinkevich FD. Detection of amitraz resistance and reduced treatment efficacy in the Varroa Mite, Varroa destructor, within commercial beekeeping operations. PLoS ONE 2020;15(1):e0227264.
[29] Rinderer TE. Resistance to the parasitic mite Varroa destructor in honey bees from far-eastern Russia. Apidologie 2001;32:381–94.
[30] Mondet F, Beaurepaire A, McAfee A, Locke B, Alaux C, Blanchard S, et al. Honey bee survival mechanisms against the parasite Varroa destructor: a systematic review of phenotypic and genomic research efforts. Int J Parasitol 2020;50(6–7):433–47.
[31] Locke B. Natural Varroa mite-surviving Apis mellifera honeybee populations. Apidologie 2016;47(3):467–82.
[32] DeGrandi-Hoffman G, Ahumada F, Danka R, Chambers M, Watkins deJong E, Hidalgo G. Population growth of Varroa destructor (Acari: Varroidae) in colonies of Russian and unselected honey bee (Hymenoptera: Apidae) stocks as related to numbers of foragers with mites. J Econ Entomol 2017;110(3):809–15.
[33] Korpela S, Aarhus A, Fries I, Hansen H. Varroa jacobsoni Oud. in cold climates: population growth, winter mortality and influence on the survival of honey bee colonies. J Apic Res 1992;31(3–4):157–64. Available from https://doi.org/10.1080/00218839.1992.11101278.
[34] DeGrandi-Hoffman G, Ahumada F, Zazueta V, Chambers M, Hidalgo G, Watkins deJong E. Population growth of Varroa destructor (Acari: Varroidae) in honey bee colonies is affected by the number of foragers with mites. Exp Appl Acarol 2016;69(1):21–34.
[35] Frey E, Schnell H, Rosenkranz P. Invasion of Varroa destructor mites into mite-free honey bee colonies under the controlled conditions of a military training area. J Apic Res 2011;50(2):138–44.
[36] Kulhanek K, Garavito A, vanEngelsdorp D. Accelerated Varroa destructor population growth in honey bee (Apis mellifera) colonies is associated with visitation from non-natal bees. Sci Rep 2021;11(1):1–15.
[37] Peck DT, Seeley TD. Mite bombs or robber lures? The roles of drifting and robbing in Varroa destructor transmission from collapsing honey bee colonies to their neighbors. PLoS ONE 2019;14(6):e0218392.
[38] Ushitani T, Perry CJ, Cheng K, Barron AB. Accelerated behavioural development changes fine-scale search behavior and spatial memory in honey bees (Apis mellifera L.). J Exp Biol 2016;219:412–8.

[39] Hoover K, Grove M, Gardner M, Hughes DP, McNeil J, Slavicek J. A gene for an extended phenotype. Science 2011;333(6048):1401. 1401.

[40] Natsopoulou ME, McMahon D, Paxton RJ. Parasites modulate within-colony activity and accelerate the temporal polyethism schedule of a social insect, the honey bee. Behav Ecol Sociobiol 2016;70(7):1019–31.

[41] Silva D, Ceballos R, Arismendi N, Dalmon A, Vargas M. Variant A of the deformed wings virus alters the olfactory sensitivity and the expression of odorant binding proteins on antennas of Apis mellifera. Insects 2021;12(10):895.

[42] Iqbal J, Mueller U. Virus infection causes specific learning deficits in honeybee foragers. Proc R Soc B 2007;274(1617):1517–21.

[43] DeGrandi-Hoffman G, Ahumada F, Graham H. Are dispersal mechanisms changing the host–parasite relationship and increasing the virulence of Varroa destructor (Mesostigmata: Varroidae) in managed honey bee (Hymenoptera: Apidae) colonies? Environ Entomol 2017;46(4):737–46.

[44] Genersch E, Von der Ohe W, Kaatz H, Schroeder A, Otten C, Büchler R, et al. The German bee monitoring project: a long term study to understand periodically high winter losses of honey bee colonies. Apidologie 2010;41(3):332–52.

[45] Seeley TD, Smith ML. Crowding honeybee colonies in apiaries can increase their vulnerability to the deadly ectoparasite Varroa destructor. Apidologie 2015;46(6):716–27.

[46] Le Conte Y, de Vaublanc G, Crauser D, Jeanne F, Rousselle JC, Bécard JM. Honey bee colonies that have survived Varroa destructor. Apidologie 2007;38(6):566–72.

[47] Seeley TD. Honey bees of the Arnot Forest: a population of feral colonies persisting with Varroa destructor in the northeastern United States. Apidologie 2007;38(1):19–29.

[48] Loftus JC, S ML, Seeley TD. How honey bee colonies survive in the wild: testing the importance of small nests and frequent swarming. PLoS ONE 2016;11(3):e0150362.

[49] Smoliński S, Langowska A, Glazaczow A. Raised seasonal temperatures reinforce autumn Varroa destructor infestation in honey bee colonies. Sci Rep 2021;11(1):22256.

[50] Rajagopalan K, DeGrandi-Hoffman G, Pruett M, Jones VP, Corby-Harris V, Pireaud J, et al. Warmer autumns and winters could reduce honey bee overwintering survival with potential risks for pollination services. Sci Rep 2024;14(1450):5410.

[51] DeGrandi-Hoffman G, Corby-Harris V, Graham H, Watkins deJong E, Chambers M, Snyder L. The survival and growth of honey bee (Hymenoptera: Apidae) colonies overwintered in cold storage: the effects of time and colony location. J Econ Entomol 2023;116(4):1078–90.

Chapter 6

Climate change as the product of a system

All outcomes are products of systems, but climate change is actually the product of two systems. The first system is the carbon cycle that is key to sustaining life on Earth. The second system is the extraction and pervasive use of fossil fuels. Multiple subsystems have been generated to optimize and utilize fossil fuels. In this chapter, I will explain the carbon cycle and how carbon levels, especially CO_2 levels in the atmospheric, are kept in check. The carbon cycle is elegant in its simplicity and ability to regulate carbon flow through all ecosystems on Earth. By regulating CO_2 levels, the carbon cycle moderates temperatures in the atmosphere and oceans that ultimately determine weather conditions.

The second system, the one that produces fossil fuels, is complicated. Numerous subsystems have been built to extract fossil fuels and use them to generate electricity, heating, support transportation, and produce agricultural chemicals and plastics. The fossil fuel subsystems are integrated into other systems of human societies, lifestyles, and economies. The combination of fossil fuel use with the carbon cycle has created a single system with climate change as its product.

6.1 The carbon cycle

We are, in the words of Joni Mitchell, "million-year-old carbon." The carbon we see today has been cycling around the planet since the Earth was formed 4.5–4.6 billion years ago. Carbon is the fourth most abundant element in the universe and is fundamental to the structure and function of all organisms. Carbon has unique properties among elements in that carbon atoms can bond together and form chains. The chains can fold in on themselves and create geometric structures such as triangles, squares, and rings. The carbon chains and geometric structures can bind together and, by adding other elements, form more complex molecules like proteins, carbohydrates, lipids, and nucleic acids like DNA and RNA (https://openstax.org/details/books/organic-chemistry). The amount of carbon on Earth is finite; only a tiny number of carbon atoms from outer space reach the Earth. All the carbon on Earth exists in rocks, sediments, the atmosphere, bodies of water, living organisms, and the remains

The Honey Bee and Climate Change. DOI: https://doi.org/10.1016/B978-0-443-28995-8.00006-4

of dead plants and animals covered deep in the Earth under layers of soil, rock, and water.

Carbon is present in all living organisms with plants harboring the greatest amounts, followed by bacteria and fungi. Carbon-based molecules flow among organisms and ecosystems through the fast and slow carbon cycles. A good place to begin a discussion of the fast carbon cycle is with photosynthesis. I discuss photosynthesis in detail in Chapter 3, but will cover it briefly here. Photosynthesis is the process where plants capture CO_2 from the atmosphere and combine it with sunlight and water to form the simple sugar, glucose.

Plants have processes to bind glucose molecules together to form complex molecules like cellulose that is used for building cell walls. The energy stored in carbon-based molecules created through photosynthesis is available to animals that consume plants (herbivores) and predators that consume herbivores (carnivores). Energy stored in carbon-based molecules is used to fuel metabolic processes such as growth and reproduction that involve building new tissues (anabolism) or digesting food and using it for energy (catabolism) (https://ugc.berkeley.edu/what-is-global-change/infographic/). Defecation and decomposition release the carbon back into the soil. Carbon is released into the atmosphere through respiration.

CO_2 is a by-product of respiration. When animals exhale, they release CO_2 into the atmosphere where it can be captured by plants and begin the cycle again. In addition to respiration, CO_2 is released into the atmosphere from events such as volcanic eruptions and forest fires (https://www.energy.gov/science/doe-explainsthe-carbon-cycle).

While the fast carbon cycle is initiated by sunlight and photosynthesis and quickly moves carbon among living organisms, the slow carbon cycle releases this element over millions of years. The slow carbon cycle begins with rain that combines with the carbon in the atmosphere to form carbonic acid. The acid slowly dissolves rocks and releases the calcium, magnesium, potassium, and sodium ions that are chemically bound in the rocks. The ions drain from the land into rivers and eventually the sea. In water, carbonic acid breaks down into bicarbonate and hydrogen. The calcium binds with bicarbonate in the ocean and forms calcium carbonate. If you have "hard water" running through the plumbing in your house, it is calcium carbonate that you see building up around your sink fixtures and faucets. Ocean life such as mollusks (e.g., oysters, clams, and mussels) crustaceans (e.g., lobsters, crabs, and shrimp), corals, and plankton use the calcium and carbonate in ocean water to make calcium carbonates to construct their shells. When the organisms die, they sink to the sea floor and over time the layers of shells and sediment are bound together to form limestone that stores the carbon. The carbon can be released through weathering of limestone rocks and begin the cycle again (https://earthobservatory.nasa.gov/features/CarbonCycle/page2.php).

About 80% of rocks containing carbon are made from the remains of shellfish. The remaining 20% are made from dead plants. When trees and plants die, especially in wet or swampy environments, they can sink beneath the soil surface and decompose slowly. The carbon in the decaying plants and trees can be sequestered underground for millennia and release carbon slowly as CO_2. In other cases, the plant matter sinks far beneath the surface and can be covered by sediments so that the carbon remains in the soil for millions of years.

The fast and slow carbon cycles are resilient systems that have kept CO_2 levels in the atmosphere in balance for millions of years despite natural disasters such a fires and volcanoes. The fast carbon cycle frees CO_2 largely through respiration, plant and animal decomposition, fire, and geological events. The slow carbon cycle sequesters carbon in the soil, oceans, and living plants and releases it gradually over expanses of time. The two cycles are interlocked systems that have maintained atmospheric CO_2 at or below 300 parts per million until recently. Human activity, primarily the burning of fossil fuels, is releasing sequestered carbon from the slow carbon cycle at levels that exceed the system's resilience. Consequently, atmospheric CO_2 is rising, leading to higher atmospheric and ocean temperatures and disruptions in the Earth's climate (for more information, go to https://earthobservatory.nasa.gov/features/CarbonCycle#:~:text=During%20photosynthesis%2/C%20plants%20absorb%20carbon,fast%20(biological)%20carbon%20cycle).

6.2 Fossil fuels: sequestered carbon

Coal, petroleum, and natural gas are fossil fuels and products of the slow carbon cycle. They are nonrenewable forms of energy because it takes millions of years to create fossil fuels. One of the most abundant fossil fuels is coal. The carbon in coal was sequestered hundreds of millions of years ago from fallen trees and plants in swampy prehistoric forests. The dead trees and plant matter sank to the bottom of the swamps, and the carbon was sequestered first as peat. As layers of peat piled up over millions of years, the pressure and heat from the weight and decay transformed the soft organic material into the sedimentary rocks we call coal. High concentrations of coal are found in underground beds or seams and reveal locations of prehistoric forests and swamps (for more information about coal formation, go to https://www.energy.gov/sites/prod/files/Elem_Coal_Studyguide.pdf).

The amount of pressure and heat that acted upon the decaying plant matter determine the hardness and carbon content in the coal. Coal formed under high pressure and heat is rock hard, has high carbon content, and burns hot. This type of coal (anthracite) is used in metal industries. Coal that is softer with lower carbon content (bituminous coal) is more abundant than anthracite and is used to generate electricity. Bituminous coal also is used to make a substance called "coke" (a fusion of carbon and ash) that burns hotter than bituminous coal. The steel industry uses coke in blast furnaces to melt iron ore into metallic iron. Two other types of coal, sub-bituminous and lignite, have relatively low amounts of carbon and high amounts of moisture. Lignite is found relatively near the soil surface, and is the most accessible type of coal. Softer coals are used mostly to generate electricity (for more information about coal and its uses, go to https://www.eia.gov/energyexplained/coal/).

Another source of sequestered carbon that formed over millions of years is crude oil (petroleum). Crude oil forms from the organic remains of prehistoric plankton, algae, and plants that lived in shallow marine environments. When the organisms died, their remains sank to the sea floor and were covered by layers of sand, silt, and rock. Over the eons, the organic material was compressed under rocks and sediments and exposed to heat from the hot mantle of the Earth. The conditions created a near total absence of oxygen and caused the decaying remains to form a waxy substance called kerogen. Heat and microorganisms broke down the kerogen so that only oil (petroleum) remained. The high heat also caused gases like methane to form and these are mined today as "natural gas."

Petroleum is found beneath the ocean floor and underground on dry land where ancient seas receded, creating dry basins (sedimentary basins). Both petroleum and natural gas tend to collect in formations underground known as oil and gas fields (for more information on petroleum and natural gas formation, go to https://personal.ems.psu.edu/~pisupati/ACSOutreach/Petroleum_2.html#:~:text=As%20temperatures%20and%20pressures%20increase,will%20remain%20trapped%20as%20kerogen). Drilling depths for oil and natural gas wells on land average f 1–2 miles but have gone as deep as 6 miles. In the ocean, wells in shallow waters average depths of about 2–3 miles but in deeper waters have gone more than 7 miles beneath the sea bed.

6.3 Reliance on fossil fuels

The earliest relationship between humans and fossil fuels goes back as far as 3600 B.P. (i.e., 1650 BCE; the Bronze Age) when inhabitants of a settlement in northwestern China used coal for cooking, heating, and melting metals [1]. There is archaeological evidence that Romans in England burned coal in the second and third centuries. In North America, Hopi Indians burned coal for cooking, heating, and hardening clay in pottery making (https://mh3wv.org/wp-content/uploads/2021/05/COAL-is-our-most-abundant-fossil-fuel-us-dept-of-energy-wv-clean-coal-tech.pdf).

Coal use continued throughout the Middle Ages across Europe and Asia, but the amount that was used was limited by its availability. Coal needed to be on the soil surface or just beneath the ground to be accessible. Coal was increasingly available throughout the Middle Ages as shafts were dug beneath the surface outcrops of coal seams (https://www2.bgs.ac.uk/mendips/minerals/coal2.html). Over time, those shafts were dug deeper, and methods to extract coal became more efficient. By the late 17th century and into the 18th century, shafts 150 m deep were being dug to extract coal. Coal mining and shipping industries continued to develop throughout the 19th century making coal readily available especially in Europe and the United States. The abundance and widespread use of coal was instrumental in fueling an era called the Industrial Revolution (https://oilprice.com/Energy/Energy-General/The-Complete-Guide-To-Fossil-Fuels.html).

The historic and current dependence on fossil fuels can be traced back to the Industrial Revolution. Between 1760 and 1840 (first Industrial Revolution) and from the 1870s to the early 20th century (second Industrial Revolution) there was a boom in mechanical engineering innovations and applications. What followed was a rapid development of machines to increase the efficiency of producing goods. During the first Industrial Revolution, factories were built and people ran machines that mass produced consumer goods, particularly textiles that before were made by hand. The invention of the cotton gin, gas street lighting, and mass steel production also occurred. Innovations that increased the efficiency of steam engines provided a power source for machinery and laid the groundwork for mass transportation by ship and rail. During the Second Industrial Revolution, there were rapid advances across many parts of society including communication (telegraph and telephone), automobile, and long-distance transportation (ships and trains). Industrial processes for mass production of steel were developed and made steel more affordable and available for construction, railroads, and industries. The use of electricity also became more widespread. Concurrent with the advances in technologies during the Second Industrial Revolution came an increasing use and dependence on fossil fuels.

Running industrial machines, generating heat and electricity for factories and homes, and producing steel have high energy demands that were met by burning coal. Coal also fueled long-distance transportation by ships and locomotives. Coal was the fuel of choice for transportation because it has three times the energy density of dry wood and requires less space for fuel storage. Mining and coal production increased rapidly from the 19th century and into the 20th century from an annual average of 8.4 million tons in 1850, to 40 million tons in 1870, and 270 million tons by 1900. Growth in coal production continues. In 2023, the United States alone produced more than 570 million tons of coal primarily to generate electrical power for heating and cooling of residential and commercial structures. Worldwide, 7800 million metric tons of coal are used per year; 63% of it to generate electricity. The remainder is used to produce iron, steel, and cement or is converted to liquid fuels (https://www.eia.gov/coal/annual/).

Although large amounts of coal are extracted and burned each year, there is even greater reliance on crude oil (petroleum). Petroleum has a long history in human societies dating back 4000 years when natural asphalt was used in Babylon as construction material. Crude oil has been used in lamps for 2000 years. Petroleum was easy to collect in some areas because the oil seeped to the soil surface through porous rock and created small puddles. More often though, the crude oil was trapped underground in reservoirs. By the

19th century, machines were developed to drill rather than dig for oil, and the availability of crude oil increased greatly. With growing availability came increased use. Oil was used in lamps as a replacement for whale oil and then for heat and as lubricants for machinery (https://guides.loc.gov/oil-and-gas-industry/history). As petroleum became more available, it replaced coal as an energy source for transportation because oil-based products generate greater power and speed at a lower cost than burning coal.

Using petroleum products for transportation rather than burning coal ushered in a new era of transportation beginning with the first liquid fueled internal combustion engine. The engine was developed by George Brayton in 1872, leading the way for Karl Benz in 1885, to patent a "vehicle powered by a gas engine" (i.e., the first automobile) (https://www.asme.org/topics-resources/content/george-brayton, https://www.invent.org/inductees/carl-benz). Automobiles continued to develop over the next 30 years, but their use was limited by their cost and unreliability. Everything changed in 1908 when Henry Ford and the Ford Motor Company produced the Ford Model-T. The vehicle, with its simple design and construction, durability, and most importantly, affordability began the era of personal transportation. Automobiles became an increasingly common mode of transportation in the early 20th century, and fueling stations and road construction increased with it. The infrastructure to support interstate transportation by trucks and automobile was greatly enhanced in mid-century by the Federal Aid Highway Act of 1956 and Highway Act of 1958. These Acts of Congress initiated by President Eisenhower provided federal funds for a national interstate highway system (https://www.archives.gov/milestone-documents/national-interstate-and-defense-highways-act).

Throughout the 20th century, transportation speeds and distances increased. Airplane development, airline industries, and eventually space exploration expanded the use of petroleum products. Currently, two-thirds of the energy used in transportation worldwide is provided by petroleum products.

Petroleum has many uses in addition to transportation. Oil is used to heat buildings and generate electricity. Many products like plastics, polyurethane, lubricants, solvents, asphalt, antifreeze, paraffin wax, polyester, rayon, nylon, pesticides, and even crayons are made by processing petroleum and natural gas (https://www.energy.gov/sites/prod/files/2019/11/f68/Products%20Made%20From%20Oil%20and%20Natural%20Gas%20Infographic.pdf).

6.4 Fossil fuels and climate change

From the Industrial Revolution to the present time, new innovations, modern conveniences, and much of our everyday life have been powered by fossil fuels. Systems are constantly being developed to extract fossil fuels and produce petroleum-based products more efficiently and at lower costs to meet the needs of a growing global population. However, fossil-fuel-based systems built to fulfill human needs run counter to the overarching system of carbon cycling that moderates atmospheric and oceanic temperatures and supports life on Earth. Seams of coal and fields of oil and natural gas created over millions of years as carbon reservoirs to maintain CO_2 levels in the atmosphere have been extracted and burned mostly over the past 300 years. Burning fossil fuels releases the sequestered carbon and raises atmospheric and oceanic temperatures because the CO_2 and other gases such as methane and water vapor absorb the sun's infrared radiation. The radiation is re-emitted back to the Earth's surface and heat is generated. The process of trapping solar radiation to generate heat is called the "greenhouse effect" because CO_2 and other gases act like the glass panels of a greenhouse to capture and retain heat (https://science.nasa.gov/climate-change/faq/what-is-the-greenhouse-effect/#:~:text=The%20greenhouse%20effect%20is%20the,that%20initially%20caused%20the%20warming).

The greenhouse effect and how different carbon-based compounds affect air temperatures is not a recent discovery. The first insights into the greenhouse effect were made in 1856. The physicist Eunice Newton Foote filled glass tubes with various gases, exposed them to the sun, and found that the heat was absorbed especially if tubes were filled with carbonic acid gas (CO_2) [2]. In 1859, John Tyndall, also a physicist, found that CO_2, methane, and volatile hydrocarbons absorb more radiant heat from the sun than other atmospheric gases. He surmised that the ability of atmospheric gases to absorb the sun's heat was important in regulating global temperature [3]. Ironically, the source of the gases Tyndall used for his experiments were from burning coal (https://earthobservatory.nasa.gov/features/Tyndall). By the end of the 19th century, the possible repercussions of burning fossil fuels and increasing atmospheric CO_2 levels were being imagined. In 1896, the chemist, Svante Arrhenius predicted that heat-absorbing gases like CO_2 would cause rises in the Earth's temperature and changes in climate [4] (to read the original publication by S. Arrhenius, go to (https://www.rsc.org/images/arrhenius1896_tcm18–173546.pdf).

Although the potential for greenhouse gases to affect global temperatures was known in the mid-19th century, the use of fossil fuels has steadily increased since the Industrial Revolution as have CO_2 levels in the atmosphere. The effects on global temperatures have been known since the 1930s and were documented by a British engineer, Guy Stewart Callendar. He collected historic weather records from 147 stations across the globe and detected significant warming (0.3°C, 32.5°F) over the previous 50 years. Callendar connected the warming to rising atmospheric CO_2 levels and correlated these with amounts of coal and other fossil fuels burned since the 1880s (https://www.rmets.org/sites/default/files/qjcallender38.pdf).

Currently, 90% of global CO_2 emissions are the products of burning fossil fuels. CO_2 emissions increased 4.5-fold during the 43-year period between 1936 and 1979 and another 1.8-fold in the following 43 years. When Charles D. Keeling began plotting CO_2 levels from the top of Mauna Loa in Hawaii in 1958, they were measured at 313 ppm [5]. Levels are updated and plotted daily (the Keeling Curve) and currently are at about 420 ppm, 182 times higher than in 1850 when they were first estimated. Current CO_2 levels are the highest in human history and causing our planet to warm faster than any time in the geological record (https://scrippsco2.ucsd.edu/history_legacy/charles_david_keeling_biography.html).

6.5 Effects of high carbon dioxide and greenhouse gas levels on the oceans

Oceans cover 71% of the Earth and hold 97% of its water. Rising CO_2 and greenhouse gas levels and the warmer air temperatures they create have effects on ocean waters and marine ecosystems. Ocean surface temperatures around the Earth have shown a consistent warming trend since the late 1950s. The rate of ocean surface warming has accelerated during the last decade (2010–19) to 0.280 ± 0.068°C per decade, 4.5 times higher than the long-term mean. Furthermore, each of the last six decades have been warmer than the previous one [6]. Oceans are warming because they absorb and store large amounts of heat from the atmosphere. The heat is transferred around the globe through ocean currents. The currents, especially at the ocean surface, regulate local temperatures and precipitation and generate weather patterns and events.

The constant motion of ocean waters both at the surface and in its depths create currents that move water around the globe. Currents can be close to shore (tidal currents), at the surface of the waters (surface currents), or in deep water below 300 m (984 ft) (deep currants). Surface currents are produced by winds that stream water horizontally. The streams can travel across the globe and be 100 m or more deep. Contributing to the formation of surface currents is the Earth's rotation (i.e., the "Coriolis Effect"). Because the Earth is spinning, it steers the direction of the wind over areas between the equator and the poles. Winds are steered to the right in the Northern Hemisphere and to the left in the Southern Hemisphere. The Coriolis Effect also deflects ocean currents to the right and left in a similar manner as with the wind. The shape of ocean basins also affects ocean currents much as the movement of water in a bowl is affected by its shape.

Surface currents work together with deep currents to circulate ocean waters around the Earth. Deep currents are produced by differences in water density and cause slow-moving large-scale circulation of ocean waters deep beneath the surface. Water density is determined by its temperature and salinity. Cool water is more dense than warm water. Also, the more salt in the water, the greater its density. Currents created by differences in water density due to temperature and salinity are called "thermohaline circulation" (https://oceanservice.noaa.gov/education/tutorial_currents/05conveyor1.html).

The circulation of ocean waters around the globe through shallow and deep currents begins when winds at the equator move warm surface water toward the North and South Poles. The warm water cools as it reaches the Poles, and its density increases. As the water freezes (sea ice), salt is released into the surrounding waters, increasing their salinity. The cold salty water sinks beneath the surface and is replaced by warmer surface waters that also will cool and sink. When cold dense surface water sinks to the ocean floor, it pushes deeper water forward and upward, causing circulation of deep ocean waters. The constant mixing of warm surface waters and colder deep waters creates a cycle of upwelling and downwelling currents and brings nutrient rich water to the surface. The nutrients support the growth of phytoplankton and seaweed that support numerous species of marine and land animals.

The circulation of ocean waters north to south in the Atlantic Ocean is called the Atlantic Meridional Overturning Circulation (AMOC). Climate change is affecting the AMOC because ocean waters are warming. Warm water is less dense than cold and does not readily sink. Furthermore, the warmer water circulating to the North Atlantic is causing Greenland's ice sheet and glaciers to melt. Glaciers and ice sheets are made from fresh water. When they melt, fresh water is added to the North Atlantic, reducing its salt concentration. The combination of warm water and lower salt concentrations reduce the amount and depth of sinking surface waters and weaken the AMOC.

Changes in AMOC will affect weather patterns, rainfall amounts, and ultimately where crops can be grown. The latest report from the International Panel on Climate Change states that the AMOC is "very likely to weaken over the 21st century" due to climate change. There is some evidence that suggests that the AMOC may already be weakening (see refs in [6]).

Loss of ice from the Greenland ice sheet and other parts of the cryosphere (parts of the Earth with permanent ice including glaciers, ice sheets, sea ice snow, and permafrost) contributes to ocean and atmospheric warming in other ways besides dilution of salt water. The cryosphere is part of a global system to moderate the global heat balance by reflecting solar radiation and radiating longwave heat. Losing parts of the cryosphere sets up a feedback loop where more heat from the ocean is released into the atmosphere through thinner sea ice and expanding areas of open water. Increasing heat causes greater loss of ice in the cryosphere, further compromising the global heat balance system needed to moderate global temperatures (see ref. [6]).

Ocean temperatures drive regional and global weather patterns because evaporation at the ocean surface causes cloud formation and precipitation. The frequency of hurricanes, typhoons, heavy rain, flooding, heat waves, and drought also are affected by ocean temperatures. Warmer oceans that melt glaciers, icebergs, and other parts of the cryosphere cause sea levels to rise and cause flooding in areas along coast lines. The rise in sea levels has more than doubled over the last three decades and contributes to dangerous storm surges, coastal erosion, and landslides (https://climate.nasa.gov/vital-signs/ocean-warming/?intent=121).

Increasing CO_2 levels in the atmosphere also are making oceans more acidic (water and CO_2 form carboxylic acid). Global surface ocean pH has declined by approximately 0.1 pH units (from 8.2 to 8.1) since the

Industrial Revolution (1770). By the end of this century (2100), ocean pH is projected to decline an additional 0.1–0.4 pH units. Global warming also has reduced oxygen solubility at the surface and decreased oxygen levels by 2% in the past five decades. The increase in acidity and decline in oxygen concentrations has impacted habitats at local and basin levels including coral reefs and other ocean ecosystems that support innumerable species of sea life (see ref. [6]). Coastal ecosystems such as mangroves, tidal marshes, and seagrass meadows that are highly efficient at sequestering carbon also are threatened.

6.6 Where do we go from here?

In the Systems chapter, I described how the rate that products are generated from a system can be in positive or negative directions. Production rates can achieve momentum that becomes increasingly difficult to reverse or to reestablish equilibrium. Systems that extract and then use fossil fuels for electricity, transportation, plastics, and agrochemical manufacturing have been generating momentum since the Industrial Revolution despite causing local pollution and global climate change. Adding to the momentum is the increase in global population and rising global gross domestic product. Essentially, more people are producing and consuming more goods produced by systems energized with fossil fuels.

Although momentum for increased energy is growing, climate change can be reversed by building two new systems. The first system is to reduce CO_2 emissions, and the second is to remove and sequester the CO_2 that fossil fuels have released into the atmosphere. Replacements for fossil fuels are well underway. Numbers of solar panels on the rooftops of homes and businesses and in large arrays on open lands have increased sharply over the past 14 years (https://seia.org/research-resources/solar-industry-research-data/). Increases in offshore wind turbines also are increasing energy as are hydroelectric power, geothermal energy, and energy generated from ocean tides [7].

A major challenge to building new systems to generate nonfossil-fuel-based energy is finding means to store the energy. Fossil fuels are essentially stored energy that is released when they are burned. Storing energy from solar, wind, and other nonfossil fuels is difficult, but subsystems for energy storage are being developed. Some energy can be stored in batteries, but their size and capacity limit their use. For vehicles, like passenger cars, the problem of energy storage is solved by recharging the batteries ideally with a nonfossil fuel energy source. However, for long-distance transportation such as aviation, maritime shipping, or long-distance trucking, new methods to store power need to be developed. Industrial processes that need high heat such as steel, glass, and cement production also present a challenge, as burning fossil fuels is currently the only means of achieving high enough temperatures to manufacture these products. However, technology for low-carbon concrete (https://www.carboncure.com/blog/concrete-corner/a-complete-guide-to-low-carbon-concrete/#:~:text=SCMs%E2%80%94which%20include%20fly%20ash, emissions%20by%20up%20to%2030%25) and green steel (https://www.weforum.org/stories/2022/07/green-steel-emissions-net-zero/) are available but prices need to be competitive before they are widely adopted.

Systems to reduce CO_2 need to include subsystems that reduce energy use. Methods to increase energy efficiency in heating, cooling, lighting, and transportation are available and are a place where each of us can contribute to reducing CO_2 levels. Lists of currently available methods to reduce CO_2 can be found at https://drawdown.org,/solutions and include recent innovations such as reformulating high carbon use products such as cement and refrigerants. Building retrofitting to improve insulation and heating and cooling efficiency along with automation systems to control thermostats, lighting, and appliances also could significantly reduce energy use and amounts of CO_2 released into the atmosphere. Homes and businesses can reduce CO_2 emissions with LED lighting, heat pumps, and low flow plumbing fixtures. Roofs designed to reflect sunlight and reduce energy for cooling can be installed into existing structures.

Improving energy efficiency and lowering emissions with new fuel formulations and vehicle designs for airplanes, ships, and trucks could significantly reduce CO_2 levels generated from long-distance transportation. CO_2 levels from local transportation could be reduced by increasing the use of electric and hybrid cars, improving thoroughfares for bicycles, and encouraging mass transportation. Designing communities for pedestrians and cyclists could reduce the practice of driving short distances and reduce CO_2 emission especially in populated areas. Using less one-time-use plastics also would reduce CO_2 levels, as plastics are petroleum-based and require high amounts of energy to produce. Problems with disposal and pollution from microplastics also would be alleviated if we used less plastic [7].

The second system that needs to be constructed is removing CO_2 from the atmosphere and storing it in a process called "sequestration." Part of this system already is in place with trees and other plants that remove CO_2 from the air and convert it to oxygen and glucose (i.e., photosynthesis). The carbon is stored in trees and plants, and in the soil when the plants die. Preserving and expanding forest land has the potential to absorb almost 10% of the CO_2 emitted yearly by fossil fuels [7]. The amount of carbon that can be sequestered depends on the composition of the forest, the types of trees that are planted, how fast they grow, and the fossil-fuel-based products they will replace when trees are harvested [8].

Preserving and reestablishing coastal and marine ecosystems also can reduce CO_2 levels because algae, mangroves, and seagrasses absorb and trap carbon as they grow. Wetlands and salt marshes capture CO_2 at four times the rate of traditional forests because much of the carbon goes several meters deep into the wet soil [9].

In addition to natural systems for reducing CO_2, direct air capture (DAC) and carbon capture technologies have been developed to remove CO_2 [10,11].

DAC removes CO_2 from the atmosphere using large turbines to suck in air and run it over filters or liquid solvers to remove the CO_2. The CO_2 is stored underground or used in industrial applications.

While DAC removes CO_2 c from the atmosphere, carbon capture technologies prevent carbon from being released in the first place. Industrial sources such as power plants are using carbon capture to lower carbon levels in their emissions. Both DAC and carbon capture technologies will be needed to lower CO_2 levels, but currently neither are efficient enough to reduce CO_2 levels if fossil fuels are still used as primary energy sources [12].

Envisioning systems that reduce carbon levels in the atmosphere and reverse climate change comes down to a simple equation; the amount of CO_2 and other greenhouse gases (e.g., methane) added to the atmosphere must be less than that removed by trees, wetlands, and carbon capture technologies. At present, the focus is on net-zero CO_2 emissions where the amount of CO_2 emitted into the atmosphere is balanced by the amount removed from it. The goal is to achieve this balance by 2050 (https://www.un.org/en/climatechange/paris-agreement).

If net-zero is achieved, the rise in global temperatures by 2050 would 1.5°C (34.7°F). However, that goal currently seems unachievable [13]. CO_2 levels are continuing to rise despite increasing use of renewable energy sources and technologies for carbon capture. The levels of CO_2 that can be removed by current technology clearly show that it is easier to add greenhouse gases to the atmosphere than to remove them. Our best hope is to limit fossil fuel use by re-imagining economies and lifestyles that incorporate the costs of CO_2 emissions. There is reticence to reducing fossil fuel use because the cost of transitioning to low carbon emission will be high and change present lifestyles. However, the consequences of climate change already are costing billions of dollars and changing lifestyles, especially in communities displaced by climate events. The 28 weather and climate disasters that occurred in the United States in 2023 totaled US$92.9 billion in damages and resulted in 492 direct and indirect fatalities (https://www.ncei.noaa.gov/access/billions/).

The current climate conditions and those predicted to occur in the future have thrust us into an "age of energy transition." We are in the early stages of transitioning from fossil fuels to renewable energy sources, and momentum of the fossil fuel system is great. However, multiple subsystems to generate energy and to remove CO_2 and other greenhouse gases can create momentum of their own, especially if there are economic benefits and improved quality of life. The momentum is already building in that global carbon emissions have started to decline, and renewable energy use is increasing rapidly (https://earth.org/momentum-is-on-our-side-in-the-fight-against-climate-change/#:~:text=Emissions%20data,of%20attaining%20the%20IPCC%20goal). The economic benefits of renewable resources will add to the momentum as renewables are less expensive than fossil fuels and less likely to experience price swings from environmental or geopolitical events. Subsystems to reverse climate change are being assembled because tools are available and new ones are being developed. However, time is not on our side and there needs to be a sense of urgency to move beyond the age of energy transition and into an age where renewable energy and carbon capture and storage are the norm.

References

[1] Qiu M, Liu R, Li X, Du L, Ruan Q, Pollard AM, et al. Earliest systematic coal exploitation for fuel extended to~ 3600 BP. Sci Adv 2023;9(30):eadh0549.

[2] Shapiro M. Eunice Newton Foote's nearly forgotten discovery. Physics Today 2021. 10.1063/PT.6.4.20210823a.

[3] Jackson R. Eunice Foote, John Tyndall and a question of priority. Notes Rec 2020;74(1):105–18.

[4] Crawford E. Arrhenius' 1896 model of the greenhouse effect in context. Ambio-Stockholm 1997;26(1):6–11.

[5] Sundquist ET, Keeling RF. The Mauna Loa carbon dioxide record: lessons for long-term earth observations. In: Mcpherson BJ, Sundquist ET, editors. Carbon Sequestration and Its Role in the Global Carbon Cycle, Volume 183. Advancing Earth and Space Sciences; 2009. p. 27–35. Available from https://doi.org/10.1029/2009GM000913.

[6] Garcia-Soto C, Cheng L, Caesar L, Schmidtko S, Jewett EB, Cheripka A, et al. An overview of ocean climate change indicators: sea surface temperature, ocean heat content, ocean pH, dissolved oxygen concentration, arctic sea ice extent, thickness and volume, sea level and strength of the AMOC (Atlantic Meridional Overturning Circulation). Front Mar Sci 2021;8:642372.

[7] Godin S. The carbon almanac: it's not too late. Penguin; 2022.

[8] Forster EJ, Healey JR, Dymind C, Styles D. Commercial afforestation can deliver effective climate change mitigation under multiple decarbonisation pathways. Nat Commun 2021;12:3831.

[9] Pendleton L, Donato DC, Murray BC, Crooks S, Jenkins WA, Sifleet S, et al. Estimating global "blue carbon" emissions from conversion and degradation of vegetated coastal ecosystems. PLoS ONE 2012;7(9):e43542.

[10] Kumar A, Madden DG, Lusi M, Chen KJ, Daniels EA, Curtin T, et al. Direct air capture of CO_2 by physisorbent materials. Angew Chem Int Ed 2015;54(48):14372–7.

[11] Okesola AA, Oyedeji AA, Abdulhamid AF, Olowo J, Ayodele BE, Alabi TW. Direct air capture: a review of carbon dioxide capture from the air. IOP Conference Series: Materials Science and Engineering 2018;413:012077. Available from https://doi.org/10.1088/1757-899X/413/1/012077.

[12] Jacobson MZ. The health and climate impacts of carbon capture and direct air capture. Energy Env Sci 2019;12:3567.

[13] Deutch J. Is net zero carbon 2050 possible? Joule 2020;4(11):2237–40.

Chapter 7

Climate change effects on colony nutrition

Increased use of fossil fuels and release of previously sequestered carbon into the atmosphere started in earnest less than 200 years ago. Organisms can adapt to changing conditions. However, there are consequences when environments change faster and more drastically than species can adapt. Species either disperse to where conditions are more amenable to their life cycle or, in extreme cases, there is a danger of extinction.

Honey bees are impacted by climate change at many levels, particularly with regard to nutrition. The nutritional needs of colonies rest solely on the availability of flowering plants. Plants establish in specific locations because conditions support their growth and reproduction. However, if factors such as temperatures and precipitation exceed plant tolerances, the stress can reduce both the attractiveness of flowers to pollinators and the nutrients provided to them in nectar and pollen. If pollination and seed set are reduced, so is the future establishment of plants and availability of food sources for pollinators.

The relationship between plants and pollinators is one of reciprocity. Plants rely on pollinators to transfer pollen and initiate seed set. Honey bees need nectar and pollen for food and to build comb. Plants make complex compounds from the sugars they create during photosynthesis and use them to attract and reward pollinators, reproduce, and protect their seeds. In this chapter, I will discuss how the nutritional needs of colonies change throughout the year and how those needs are met by seasonal plants.

Flowering plants and pollinators are interdependent, and either thrive or struggle together. Under favorable conditions, the mutualism between plants and pollinators enabled both to establish populations and expand across landscapes. However, climate change, especially effects on temperature and rainfall, is affecting plant–pollinator relationships by limiting flower abundance and attractiveness, reducing the availability nectar and pollen, and creating nutritional stress on colonies that impacts their health and growth.

7.1 What is nutrition?

Before discussing honey bee nutrition, it is best to define "nutrition" and its relationship to health in a general sense. Nutrition is a process where components

The Honey Bee and Climate Change. DOI: https://doi.org/10.1016/B978-0-443-28995-8.00015-5

of food (i.e., nutrients) are broken down into simple molecules (i.e., digested) and used in processes that produce energy, build tissues, and generate immune responses. Collectively, the processes are referred to metabolism. From a systems perspective, nutrients in food are the raw materials that are absorbed into the circulatory system and transported into multiple metabolic subsystems located in the cells of various organs. The subsystems run continuously, and their interactions enable the organism to function and respond to its environment.

Nutrients in food are classified as carbohydrates, proteins, lipids (fats), vitamins, minerals, and water. Carbohydrates are sugars, starches, and fiber. Honey bees obtain carbohydrates from nectar. Pollen also contains complex carbohydrates such as cellulose, pectin, and sporopollenin on the outer surface of pollen grains. Carbohydrates are animals' main source of energy for movement, cognitive function, and responding to environmental conditions. In animals, excess carbohydrates are stored as glycogen and fat.

Proteins are large, complex, three-dimensional organic (i.e., they contain carbon) molecules used to build tissues, form enzymes that increase the speed (i.e., catalyze) of reactions, transport molecules in and out of cells, and produce energy if carbohydrate levels are low. Proteins are made of chains of amino acids. The three-dimensional structure of proteins is determined by the sequence of amino acids. Some amino acids can be produced in the body by reassembling molecules obtained from food. These amino acids are called, "nonessential." Amino acids that cannot be made in the body and must come from food are "essential amino acids." Ensuring that diets contain all the essential amino acids is critical for metabolism because if they are missing, certain proteins might not have the structure needed to be functional. In honey bees, protein comes primarily from pollen.

Lipids are large organic molecules that are used for energy storage (fats), absorption of fat-soluble vitamins (e.g., A, D, E, and K), and synthesis of cell membranes, oils, steroids, and wax. Lipids are composed of fatty acids chemically bonded to glycerol. For example, if three fatty acid molecules bind to a glycerol molecule, they create a type of lipid called a triglyceride. Like essential and nonessential amino acids, certain fatty acids can be synthesized by bees (nonessential), while others need to come from their diet (essential). Omega-3 (alpha linolenic acid) and Omega-6 (linoleic acid) are essential fatty acids that in honey bees play a role in brood development, adult lifespan, immune response, and learning [1,2]. The dietary source of lipids for honey bees is pollen. Amounts of essential and nonessential fatty acids in the pollen depend on the plant source (see ref. [3]).

Vitamins are organic molecules present in relatively small amounts in food. Vitamins are necessary for the functioning of enzymes in biochemical reactions in cells. The reactions are responsible for energy production, metabolism, and DNA repair. Minerals are important components of diet because they affect muscle, nerve, and immune function, are components of hormones and enzymes, and are important in water regulation. Honey bees obtain vitamins and minerals from pollen but also from water sources (see ref. [4]).

7.2 The role of microbes in nutrition

Physiological systems that convert nutrients to energy, support growth and development, and generate immune responses do not work alone. There are systems within the systems, and they are composed of microbes (e.g., bacteria and fungi). Microbial communities (i.e., the microbiota) find places to establish throughout the body of their hosts. Microbes can interact with their host by exchanging resources or preventing the growth of other microbes that might be pathogenic. Many microbes and their hosts have a mutualistic relationship. The microbes rely on the host for food, shelter, and a means of dispersal. The host benefits from the microbes, as they can perform or augment physiological processes not included in host's genetic blueprint. In honey bees, microbes live in the bees, their food stores, and the hive itself.

Although microbial communities can establish in numerous places throughout the body, studies with the honey bee microbiota have focused on the gut and food stores (see refs [5,6]). The microbiota in the gut of worker honey bees is dominated by five lineages of bacteria. This core bacteria seem to have evolved with the bees about 80 million years ago [7]. Each lineage has a common ancestor and distinct genetic or phenotypic characteristics. In addition to the bacteria, some fungal species also are present in the gut, but their presence is not consistent, suggesting that they might be transient residents [8].

The composition of the gut microbiota in honey bees (and in other organisms) is fundamental for maintaining health. In honey bees, microbes in the gut aid in digestion, especially of pollen. Microbes also metabolize sugars and potentially detoxify plant compounds in nectar. The microbes also stimulate immune responses and provide protection against other microbes that could be pathogenic. The microbiota also play important roles in metabolic pathways in worker bees that regulate feeding behaviors, olfactory learning, and memory (see ref. [6]).

The gut microbiota are extensions of the host's digestive, detoxification, and immune systems and augment their capabilities. For example, in honey bees, pollen is the major source of protein, lipids, vitamin and minerals. However, those nutrients are encased in a coat comprised of complex sugars (i.e., polysaccharides) that cover the pollen grain. Honey bees do not have the enzymes to digest polysaccharides, but instead rely on core gut microbes [9]. The microbes produce short-chain fatty acids from the polysaccharides that bees can use as sources of nutrients.

In addition to producing fatty acids, gut microbiota might also provide amino acids to bees. Microbes have biochemical pathways that synthesize amino acids from compounds supplied by their host. In worker bees with and without midgut microbiota, those with core microbes had higher amino acid levels in the hindgut and hemolymph (see ref. [6]). The addition of amino acids by gut microbiota could improve nutrition in honey bees especially when protein in their diet is limited.

Microbes can extend their host's detoxification systems. Heavy metals and toxic compounds can be translocated into plants from contaminated soils. Plants also can produce toxic compounds to deter feeding by insects and other herbivores. The core microbiota of honey bees and bumblebees have genes for heavy metal transport and sequestration and might provide some resistance to the negative effects these molecules might have on bees [10]. Gut microbes might also detoxify compounds such as amygdalin, a potentially toxic cyanide-containing compound occurring in almond pollen. Bees can be exposed to this compound when collecting almond nectar and pollen, although bees rarely show negative effects. A study to determine if honey bees can detoxify amygdalin found that while they can partially degrade the compound, complete degradation requires specific strains of bacteria present in the gut core microbiota [11]. Evidence for gut microbes detoxifying other compounds that might be harmful to bees, however, is limited. Interestingly however, honey bees and bumble bee have fewer genes to detoxify plant toxins than other insects [12], suggesting that those duties might be delegated to microbes that reside in their digestive system.

Resident microbes can extend the host immune system by preventing the growth of pathogens and parasites. The gut microbiota of honey bees can protect against potentially pathogenic bacteria, fungi, and possibly mRNA viruses (see ref. [6]). The gut microbiota also can increase the expression of immunity genes, although how this is accomplished is still unclear.

7.3 Microbes in bee bread

In addition to microbes within an organism, communities of microbes also can preserve food stores especially if the foods are fermented. In honey bees, pollen is stored in comb cells as bee bread. When pollen collecting honey bees return from the field, they drop off their pollen loads into comb cells where it is mixed with honey, glandular secretions, and enzymes and packed into the cells. Cells with bee bread contain a layered mixture of pollens collected by bees foraging on different plants.

Bee bread is made through the process of fermentation where microbes break down complex sugars surrounding the pollen grains. During the fermentation process, the richness of microbial communities present when bee bread is first made is reduced to a smaller relatively stable population that includes bacteria common to the core microbiome in the gut. In this way, bee bread is not just a food source but also a reservoir for gut microbes. In addition to bacteria, fungi that could support pollen preservation and inhibit the growth of pathogens also are present [13–15].

The action of microbes in bee bread might provide a predigestion process for the pollen by breaking down the complex sugars on the outer coating of the pollen grain. If predigestion occurs, then the nutrient composition in bee bread might differ from pollen. There is evidence that the nutrients in pollen and the bee bread made from it can differ. A study was conducted where colonies of bees collected pollen and converted it to bee bread. Analysis of the pollen and bee bread indicated that the bee bread was more acidic than the pollen and had lower concentrations of soluble proteins. The bee bread, however, had higher concentrations of all essential amino acids except tryptophan [16]. Although pollen often is evaluated for its nutritional value, what might be a more accurate evaluation is the nutrients available after it had been converted to bee bread.

7.4 Supplemental feeding effects on gut microbiota

There are instances where flowering plants are not available, so beekeepers feed pollen substitutes and sugar solutions to their colonies. If diet affects the composition of gut microbiota, what happens to bees when fed artificial diets? Several studies show that supplemental feeding can negatively affect the honey bee gut microbiota and increase susceptibility to pathogens. For example, feeding colonies sucrose solution lowers the abundance of core gut species in relation to potentially pathogenic species (e.g., *Serratia* spp. and *Nosema ceranae*) in bee guts [17]. Feeding protein supplements containing soy protein or casein, which are not natural parts of a honey bee's diet, also can cause disruption of core gut communities and higher incidence of bacteria such as *Streptococcus* spp. and *Staphylococcus* spp., which are not normally found in guts of healthy worker bees [18].

If the gut microbiota is disrupted due to lack of pollen, pesticides exposure, or feeding artificial diets, can it be restored by feeding probiotics? There are several commercially available probiotic formulations available to beekeepers, but most consist of non-native microorganisms, including bacteria and fungi from the food industry. Studies indicate that the probiotics do not colonize in bees or improve their health (see ref. [6]). One method that does appear to reestablish the gut microbiome after a disruption is by feeding bee bread to the colony [17].

7.5 Seasonal nutrition requirements

Previously, I mentioned that to understand honey bees, you need to hold two thoughts in your mind simultaneously; honey bees as individuals and as a collective in the colony. When considering worker bee nutrition, two additional considerations are needed. These are the age of the individual worker and the time of year. Nutritional needs of all organisms depend on their stage of life. Young individuals that are growing and developing toward reproduction differ in metabolic demands from adults. In honey bees, this applies not just to individuals but to the colony at large.

Colonies are similar to young individuals in the spring and summer when they are growing and expanding and to adults in late summer and fall when brood rearing is declining. Honey bees are herbivores; that is, they obtain their nutrition from plants. Herbivores provide a good model for nutritional needs of honey bees and how they are met by the changes in the nutritional landscape.

The annual cycle of many herbivores begins in the spring with reproduction and caring for the young. Growth and development occur throughout the summer. In the fall, food is collected and stored for the winter. Landscapes have seasonal dynamics that provide nutrients to herbivores that support their annual cycle. In the spring, nutrient-dense foods such as plant shoots meet the high nutritional demands of pregnancy and lactation. Staple resources like stems and leaves provide the bulk of the diet during the summer when growth and development occur. In the fall, lipid-rich seeds and fruit supply high-calorie foods that are stored in fat tissue or cached in cavities and burrows to use when plants lie dormant in winter [19].

Honey bees follow the blueprint of other herbivores but with modifications based on physiology and lifestyle. Changing nutritional needs of many herbivores are met by consuming different parts of plants made available to them during the year. Honey bees feed only on pollen and nectar. However, the nutrients in pollen can differ with the seasonal progression of plants across landscapes and support the colony's changing nutritional needs. Studies analyzing the nutrients in pollens collected by honey bees at different times of year found that pollen from flowers blooming in spring and early summer have higher concentrations of amino acids and fatty acids that support brood care compared with pollens collected in the fall [20–22]. In contrast, fall pollens can have higher levels of nutrients needed for overwintering. For example, amino acids that support the transport of nutrients from the fat body to flight muscles are higher in fall than in spring pollens [20]. Flight muscles need a steady supply of nutrients during winter, as they are used to generate heat in the winter cluster.

Considering honey bee colonies as having dynamic season-based nutritional requirements weaves together colony population growth with flowering plants across time in the landscapes where honey bees forage. Unlike herbivores that can migrate to more abundant food sources during the year, honey bee hives, unless moved by a beekeeper, are stationery and forage within about 1–2 km (0.6–1.2 miles) radius depending on the availability of flowering plants [23]. Therefore, to meet seasonal nutritional requirements, landscapes need a diversity of plants that flower sequentially from spring to fall and favorable environmental conditions that allow them to flourish.

7.6 Production of nectar

Earlier, we discussed how plants make glucose from water, CO_2, and sunlight through the process of photosynthesis. Plants store glucose by linking the molecules together to form long chains called starches. Starches can have complex linear or branched arrangements of glucose molecules. When plants need energy, they break down starch granules into glucose, fructose, and sucrose molecules.

There are two categories of starches: transitory and storage starches. Starch that plants make in leaves is transitory because it is degraded at night to provide energy for metabolism and growth. Storage starches are in tissues where photosynthesis does not occur, such as stems, roots, tubers (i.e., potatoes), and seeds. Those starches supply energy when trees and other perennial plants are dormant in winter and as they grow in the spring. Seed germination also is fueled with energy from stored starches [24].

Another place where starches can be stored is in glands called nectaries that produce and secrete nectar. The location of nectaries depends on the shape of the flower, but they usually are positioned at the flower base near the stamens and pistils. The nectaries are positioned to increase the chances that pollination will occur when bees collect nectar.

Nectaries are comprised of two types of cells: thin-walled epidermal cells on the outer surface and cells beneath the epidermal layer, called parenchyma cells that contain starch granules. The sugars in nectar come from breaking down the starch granules into glucose, fructose, and sucrose in the parenchyma cells. The high sugar concentration in the cells triggers an influx of water from the plant's circulatory system (i.e., phloem). The water can contain amino acids, minerals, secondary plant compounds, ions such as potassium, and traces of lipids that are carried by the phloem into the nectaries (see refs [25,26]). The flow of water into the cell increases the hydrostatic pressure, causing nectar to exude through pores to the surface of the nectaries (see ref. [27]).

The concentration of sugar and other compounds in nectar is affected by the shape of the flower. Nectar in dish-shaped flowers like those on apple or cherry trees is more exposed to weather conditions. High humidity can dilute the nectar, or warm windy conditions can concentrate it. Amounts of nectar also can differ among flowers on the same plant or inflorescence. For example, there can be declining amounts of nectar from lower to upper flowers on stalks. A nectar gradient is created that encourages foraging bees to move upward while collecting nectar on the stalk and in the process transfer pollen among the flowers [28].

When pollinators visit flowers, they remove the nectar at the surface. The nectaries can continue to secrete nectar if sugars and water are available, causing flowers to receive multiple visits that increase the chances of pollination. Nectar can contain 10%–70% sugar, but the most attractive nectars to honey bees have sugar concentrations between 30% and 50% and contain sucrose rather than glucose or fructose [29]. The amino acids in nectar, although present at very low concentrations, have multiple functions. They might help attract foragers to the flowers because amino acids provide odor and flavor that stimulate bees to consume the nectar. The mixture of amino acids can make the nectar distinctive when scout bees return to the hive and recruit new foragers. Foraging can be affected by other compounds in the nectar. Flowers of coffee plants secrete nectar with caffeine. This compound has profound effects on long-term memory in honey bees and may help foragers remember the nectar source [30].

In addition to the sugars and other nutrients in nectar, the solution also contains microbes (see refs [31,32]). The microbes (e.g., yeasts and bacteria) in nectar can change sugar concentrations. They also affect the odors emanating from flowers. Honey bees can detect volatile compounds produced by microbes and the odors can influence pollinator preferences [33,34]. Microbes also can influence the taste of nectar, thereby influencing the attractiveness of the flowers [35]. The bacteria collected with nectar are found in the alimentary tract of honey bees and in the hive environment [36].

7.7 Pollen

Plants build many complex structures with the products of photosynthesis; petals of flowers, wood and bark, leaves, and roots. Of unmatched complexity and elegance is the structure of pollen (Fig. 7.1). Individual pollen grains contain the male genetic material of the plant. To protect the genetic information from the environment, pollen grains have a durable outer layer (exine) comprised of a remarkable substance called sporopollenin. Referred to as the "diamond of the plant world," sporopollenin is among the most resilient organic materials on Earth [37]. The durability of sporopollenin is due to its molecular structure of multiple cross-linked chains of carbon, hydrogen, and oxygen that form an extraordinarily tough biopolymer (multiple chains of molecules that are chemically bound together). An indication of the toughness and stability of sporopollenin is that the compound has been found to be chemically intact on pollen grains recovered from 500-million-year-old sedimentary rocks (Fig. 7.1).

Beneath the exine is the inner layer of the pollen grain called the intine. The intine is made of cellulose and pectin and is similar in structure to a plant cell wall. The intine surrounds the space containing the cell that houses the male gametes (i.e., a reproductive cell) and the cell that forms the pollen tube. When pollen is deposited on the stigma, the pollen tube grows out of the pollen grain through pores in the intine and down the style to the ovary of the pistil where fertilization occurs. Each pollen grain usually contains two male gametes. One of the gametes fuses with the female gamete in an ovule of an ovary to form an embryo. The second male gamete shapes a nutritional tissue called the endosperm that protects the embryo, provides it with nutrients from starches and proteins, and controls the embryo's growth into a seed [38].

Pollen grains are produced in a cavity within the anther called the pollen sac. Each anther usually contains four pollen sacs that split open to release pollen. Pollen grains of plants that rely on pollination by bees or other animals

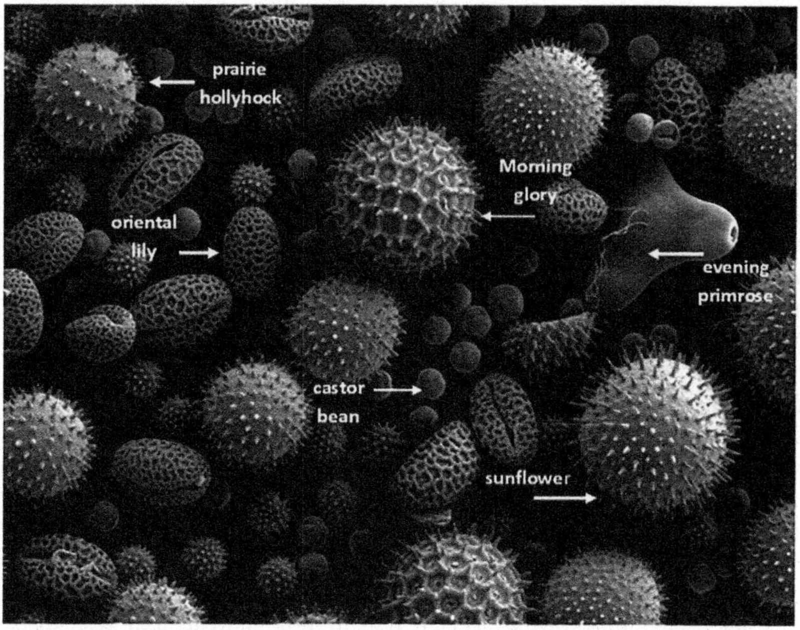

FIGURE 7.1 Scanning electron microscopy image of pollen grains. Scanning electron microscopy images of pollen grains from sunflower (*Helianthusannuus*), morning glory (*Ipomoea purpurea*), prairie hollyhock (*Sidalcea malviflora*), oriental lily (*Lilium auratum*), evening primrose (*Oenothera fruticosa*), and castor bean (*Ricinus communis*). Each plant species produces pollen grains with distinctive shapes and textures.
The image was produced by NASA/Goddard Space Flight Center.

are not shed from the anthers as single grains as are most wind pollinated plants, but as clumps of pollen stuck together by a lipid-rich sticky substance called the pollenkitt. Pollen grains adhere to the branched hairs of honey bees due to the stickiness of the pollenkitt [39]. The pollenkitt can contain compounds that produce species-specific odors and chemical cues to help pollinators find the pollen in a flower [40].

7.8 Nutrients in pollen

The compounds plants make to protect pollen grains and support reproduction have been co-opted by bees for their own nutritional needs. Pollen is the primary source of protein, nitrogen, amino acids, lipids, fatty acids, essential sterols, minerals, and vitamins for honey bees [29,41]. Colonies need large amounts of pollen to rear brood and maintain growth and will collect 55–100 lbs (24–45 kg) annually. Worker bees that are rearing brood need their diet to contain 20%–25% protein. About 25–37.5 mg of protein are needed to rear a single larva [22,42].

Pollen varies in nutrient composition depending on the plant species. Concentrations of protein are usually between 2.5% and 61%. Wind-pollinated plants such as pines (*Pinus* spp.), junipers (*Juniperus* spp.), and redwoods (*Sequoia* spp.) have pollen with the lowest protein content. Pollen from primrose (*Primula* spp.) and nightshades (*Solanum* spp., e.g., peppers, tomatoes, and eggplant) have the highest protein concentrations.

Although protein concentration is important for assessing the pollen's nutritional value, of greater significance is the bee's ability to digest the pollen and access the nutrients it contains. Studies measuring protein digestion in honey bees found that depending on season and pollen type, about 20%–80% of the soluble protein in pollen is digested by honey bees [20,43].

Proteins are made by binding amino acids together into long chains and folding them chemically into three-dimensional structures. The amino acids in protein molecules become accessible when bees break down (i.e., digest) the protein in pollen grains to their amino acid components. The most nutritious pollens contain high concentrations of essential amino acids. Essential amino acids play critical roles in the assembly of proteins bees use to support egg production, growth and development of larvae and adult bees, queen and worker jelly production in the hypopharyngeal glands, digestion, immunity, and other vital functions. Although pollen from individual flower species might not contain all the essential amino acids honey bees need, the combinations of pollens brought into the hive by foragers collecting pollen from a variety of plants increase the chances that all essential amino acids will be available to the colony [29].

In addition to protein, pollen is rich in lipids (2%–20% by dry mass). Lipids are fatty, oily, or waxy substances. Some lipids are part of the outer surface of the pollen grain (pollenkitt) and others are within the grain itself. The lipid portion of pollen and its fatty acid composition depend on the plant species that produce them. Like certain amino acids, there are essential fatty acids that bees cannot synthesize but need to obtain from pollen. Of these, the most important are linoleic (Omega-6) and alpha linolenic acid (Omega-3). Essential fatty acids must come from food sources and play key roles in cognitive function and learning. Other fatty acids have antimicrobial properties and are part of the colony-level immunity (see ref. [29]). Lipids and the fatty acids they are composed of also serve as energy sources for development and are components of cell membranes, waxes, and oils that waterproof the exoskeleton and the beeswax used to construct comb cells.

Pollen provides bees with other essential nutrients that are critical to metabolism. For example, certain lipids called sterols that honey bees cannot synthesize on their own are present in pollen. The sterols are needed for ovary maturation, stress responses, and development of larvae into pupae and adults, (see ref. [44]). Pollen also can contain vitamin C, vitamins B, and fat-soluble vitamins A, D, E, and K needed for brood rearing [45]. Finally, bees obtain minerals from pollen and use them in metabolic processes, tissue structure, nerve impulses, muscle contraction, and the exchange of water and solutes across cells and tissues.

7.9 Effects of heat waves and drought on flowering and nectar and pollen production

Floral displays are connecting points between systems supporting plant reproduction and pollinator nutrition. Honey bees and the plants they forage establish in areas with favorable ranges of environmental conditions. However, environmental conditions that expose plants to conditions beyond those ranges are occurring with greater frequency. In North America, the period when heat waves occur is now 40 days longer than it was in the 1960s (US Global Change Research Program; https://www.globalchange.gov/) [46]. The occurrence and intensity of droughts are predicted to increase with climate change across many regions worldwide [47,48]. While the vegetative tissue of plants (e.g., leaves and stems) might show resilience to environmental stress, flowers are more vulnerable as seen by reduced yields in agricultural crops when high temperatures, drought, or cold occurs during the flowering season (see ref. [46]).

The effects that heat and drought have on flowering begin with the stress they cause on photosynthetic tissue. During drought and heat waves, leaves can reach temperatures where rates of photosynthesis diminish (> 34°C, 93°F) reducing the plant's ability to produce glucose (see ref. [49]). Plants respond to heat or drought with stunted growth and fewer flowers [50,51]. Flower number and density are key factors affecting the attractiveness to pollinators (see ref. [49]). The shape, color, and scent of flowers also can be affected, making the flowers more difficult for pollinators to find [46,52].

Stress on photosynthetic tissues from heat and drought affect the quantity and quality of nectar. The sugars in nectar are products of photosynthesis [53]. During heat waves, starch granules in nectary cells are smaller as are amounts of glucose and other sugars in nectar. Production of amino acid, protein, fatty acid, lipid, and sterol also depends on photosynthesis, and they are present in lower amounts in nectar during heat waves.

The consistency of nectar is affected by high temperatures and drought because they lower soil moisture. There is less water flowing into nectary cells, causing the little nectar that is produced to have more concentrated sugar and amino acid composition. The nectar also has higher viscosity, making it difficult for honey bees and other pollinators to collect [54]. High temperatures might also affect the microbial communities in nectar, changing the sugar concentration and odors emanating from the nectaries. The changes in nectar composition and odors that serve as cues for pollinators could change foraging behaviors and the likelihood of pollination [55].

Heat waves and drought also affect pollen production. Developing pollen grains obtain nutrients from the plant's circulatory system (i.e., the phloem).

The nutrients are transported to a layer of nutritive cells and then to the developing pollen grains [56]. Heat stress and drought disrupt the transport system and alter the amounts of protein, lipids, and amino acids that reach the developing pollen grains. Less pollen is produced, and the nutrient composition is diminished [52,57–59].

Less nutritious pollen limits colony growth especially when concentrations of essential amino acids are reduced. A study on the effects of high temperatures on pollen from Borage (starflower) plants (*Borago officinalis*) showed reduced concentrations of the essential amino acid, isoleucine [60]. Lower levels of this amino acid could reduce brood rearing because honey bee larvae release isoleucine to communicate with nurse bees when they need food [61–63]. Borage is highly attractive to honey bees and other pollinators so reduced isoleucine levels could affect many pollinating species.

High temperatures can decrease levels of nutrients in pollen even if those temperatures occur before flowers bloom. Buds on blueberry (*Vaccinium corymbosum* L) plants exposed to 37.5°C (99.5°F) developed flowers that produced pollen with reduced protein and amino acid concentrations. The reductions were severe enough that Blue Orchard Mason Bee (*Osmia lignaria*) larvae could not complete their development into adults when they were reared on the pollen [59,64].

7.10 Effects of high CO_2 levels

CO_2 levels have been rising since the early 20th century, with the greatest increases occurring over the last 60 years (https://www.climate.gov/media/12990). High CO_2 levels impact flowering and the nutrients in nectar and pollen. When CO_2 levels are high in plant tissue where photosynthesis occurs (e.g., leaves), nitrogen levels decline. Nitrogen is needed to produce amino acids, so low levels decrease protein synthesis in many species of plants. The exception is legumes because these forbs fix nitrogen through a symbiotic relationship with Rhizobium bacteria living in the root nodules of the plant [65].

Reduced protein levels in some plants may have been occurring for decades. A study comparing protein concentrations in Goldenrod (*Solidago* spp.) pollen collected across the United States and Canada from 1842 to 1998 and from Maryland, USA (2008, 2012, and 2014) and Texas, USA (2012 and 2014) showed a steady decline in pollen protein, with an overall decrease of about 33%. The sampling period captures the increase in CO_2 occurring at the onset of the industrial revolution through the 20th and into the 21st century. The decline in pollen protein was highly correlated with a rise in CO_2 levels. The sharpest decline in protein occurred during the latter half of the 20th and early 21st century when CO_2 rose from about 315 ppm in 1960 to 398 ppm in 2014 [66]. In May of 2025, the global average for CO_2 set a new record high at

430.5 ppm (https://www.climate.gov/news-features/understanding-climate/climate-change-atmospheric-carbon-dioxide).

Although protein levels decline with rising CO_2 levels, carbohydrate production increases, and this can stimulate vegetative growth. In some plants, increases in vegetative growth causes a delay in flowering. In others, bloom can occur earlier because plant growth is accelerated (see ref. [46]). Higher production of carbohydrates can cause some plants (e.g., forbs) to produce bigger flowers that bloom earlier and secrete more nectar (e.g., fireweed, *Epilobium angustifolium*) [67]. However, amounts of sugar and amino acids in the nectar can be reduced.

7.11 False springs

In March of 2007, the Midwest, Southeast and mid-Atlantic portion of the United States experienced near-record warmth that stimulated growth in crops and vegetation. In early April, arctic air flowed into the regions, causing freezing temperatures and widespread agricultural losses and damage to plants in unmanaged landscapes [68]. Unseasonably warm temperatures in late winter or early spring followed by freezes are called false springs.

Unusually warm temperatures early in the year are occurring more frequently, particularly at higher elevations. In the western United States, snowpack influences temperature and soil moisture. As temperatures are becoming warmer, more precipitation is falling in the form of rain instead of snow. The snow melts sooner, causing warmer temperatures to occur earlier. The warming changes the timing of when plants begin to flower in spring and summer (see refs [69,70]). If spring frosts occur when leaf and flower buds are beginning to form or plants have begun leaf or flower emergence, they are vulnerable to frost damage. Spring frosts limit flowers during the period when honey bee colonies need pollen to build populations after overwintering. Damage to flowers also reduces seed set and the availability of pollen sources in subsequent years.

7.12 Conclusions

The effects of climate change on honey bee nutrition and health might be one of the most difficult causes of colony losses to trace. The effects can be deceptive because flowers are available and foragers are collecting nectar and pollen. However, if essential nutrients are missing, the colony will not grow and will be more vulnerable to diseases. Should the colony die, it could appear that the queen was failing or, if disease is evident, that pathogens were responsible. However, the root cause of the colony loss could be poor nutrition due to climate change.

The consequences of poor nutrition on colony growth and survival might not be realized for several months after they occur. As colonies come out of winter, false springs can damage flowering plants and limit pollen and nectar when populations are at their lowest point of the year. The trajectory for colony growth in the summer and fall is set in the spring. If poor nutrition lowers the trajectory, the consequences will be felt throughout the year.

Colonies need to grow during the summer so that they are large enough to survive the winter. Summer heat waves and drought that limit resources and reduce colony growth set the stage for colony losses over the winter. In the fall, colony growth slows and bees store resources to survive periods of confinement in the hive. The resources also are needed to rear brood before foraging resumes in the spring. Environmental conditions that reduce amounts of resources stored in colonies jeopardize survival over the winter and early spring.

The relationship between flowering plants and pollinators creates a feedback loop between them. Like all feedback loops, momentum can be created in positive or negative directions. When flowering plants are plentiful and rewarding, pollinators are attracted to them and collect nectar and pollen. Pollinator populations grow and pollination and seed set occur ensuring that flowering plants will exist in the future. If conditions are favorable in subsequent years, populations of flowering plants and pollinators can continue to grow and create momentum in a positive direction. However, when environmental conditions limit flowering plants and the nutrients they provide, foraging and pollination decline and so does seed set. Limited resources also reduce the growth of pollinator populations. In this scenario, momentum is moving in a negative direction, causing there to be fewer plants and pollinators each year.

Expanding landscapes planted as forage for pollinators can improve nutrition but only if the plant responses to environmental effects from climate change are known. Currently, the impact of climate change on flowering and nutrients in pollen and nectar have been described for only a few of the plants that bees forage. Studies are needed on many other plants that serve as nectar and pollen sources throughout the year and find those that are resilient to stress from heat, drought, high CO_2, and frost tolerance and other repercussions of false springs.

References

[1] Arrese EL, Soulages JL. Insect fat body: energy, metabolism, and regulation. Ann Rev Entomol 2010;55:207–25.

[2] Arien Y, Dag A, Zarchin S, Masci T, Shafir S. Omega-3 deficiency impairs honey bee learning. Proc Natl Acad Sci U S A 2015;112(51):15761–6.

[3] Lau P, Lesne P, Grebenok RJ, Rangel J, Behmer ST. Assessing pollen nutrient content: a unifying approach for the study of bee nutritional ecology. Philos Trans R Soc Lond B Biol Sci 2022;377(1853):20210510.

[4] Ansaloni LS, Kristl J, Domingues CE, Gregorc A. An overview of the nutritional requirements of honey bees (Apis mellifera Linnaeus, 1758). Insects 2025;16(1):97.

[5] Raymann K, Moran NA. The role of the gut microbiome in health and disease of adult honey bee workers. Curr Opin Insect Sci 2018;26:97–104.

[6] Motta EVS, Moran NA. The honeybee microbiota and its impact on health and disease. Nat Rev Microbiol 2024;22:122–37.

[7] Kwong WK, Medina LA, Koch H, Sing KW, Soh EJY, Ascher JS, et al. Dynamic microbiome evolution in social bees. Sci Adv 2017(3):e1600513.

[8] Decker LE, San Juan PA, Warren ML, Duckworth CE, Gao C, Fukami T. Higher variability in fungi compared to bacteria in the foraging honey bee gut. Microb Ecol 2023;85(1):330–4.

[9] Zheng H, Perreau J, Powell JE, Han B, Zhang Z, Kwong WK, et al. Division of labor in honey bee gut microbiota for plant polysaccharide digestion. Proc Natl Acad Sci U S A 2019;116(51):25909–16.

[10] Rothman JA, Leger L, Graystock P, Russell K, McFrederick QS. The bumble bee microbiome increases survival of bees exposed to selenate toxicity. Environ Microbiol 2019;21(9):3417–29.

[11] Motta EV, Gage A, Smith TE, Blake KJ, Kwong WK, Riddington IM, et al. Host–microbiome metabolism of a plant toxin in bees. Elife 2022;11:e82595.

[12] Berenbaum MR, Johnson RM. Xenobiotic detoxification pathways in honey bees. Curr Opin Insect Sci 2015;10:51–8.

[13] Disayathanoowat T, Li H, Supapimon N, Suwannarach N, Lumyong S, Chantawannakul P, et al. Different dynamics of bacterial and fungal communities in hive-stored bee bread and their possible roles: a case study from two commercial honey bees in China. Microorganisms 2020;8(2):264.

[14] Wang Y, Ma L, Xu B. Bee wisdom: exploring bee control strategies for food microflora by comparing the physicochemical characteristics and microbial composition of beebread. Microbiol Spectr 2023;11(6):e01818-23.

[15] Ghosh S, Namin SM, Jung C. Differential bacterial community of bee bread and bee pollen revealed by 16s rRNA high-throughput sequencing. Insects 2022;13(10):863.

[16] DeGrandi-Hoffman G, Eckholm BJ, Huang MH. A comparison of bee bread made by Africanized and European honey bees (Apis mellifera) and its effects on hemolymph protein titers. Apidologie 2013;44(1):52–63.

[17] Huang SK, Ye KT, Huang WF, Ying BH, Su X, Lin LH, et al. Influence of feeding type and Nosema ceranae infection on the gut microbiota of Apis cerana workers. Msystems 2018;3(6):e00177.

[18] Powell JE, Lau P, Rangel J, Arnott R, De Jong T, Moran NA. The microbiome and gene expression of honey bee workers are affected by a diet containing pollen substitutes. PLoS ONE 2023;18(5):e0286070.

[19] Owen-Smith N, Cooper SM. Nutritional ecology of a browsing ruminant, the kudu (Tragelaphus strepsiceros), through the seasonal cycle. J Zool 1989;219(1):29–43.

[20] DeGrandi-Hoffman G, Gage SL, Corby-Harris V, Carroll M, Chambers M, Graham H, et al. Connecting the nutrient composition of seasonal pollens with changing nutritional needs of honey bee (Apis mellifera L.) colonies. J Insect Physiol 2018;109(June):114–24.

[21] Donkersley P, Rhodes G, Pickup RW, Jones KC, Wilson K. Honeybee nutrition is linked to landscape composition. Ecol Evol 2014;4(21):4195–206.

[22] Quinlan G, Milbrath M, Otto C, Smart A, Iwanowicz D, Cornman RS, et al. Honey bee foraged pollen reveals temporal changes in pollen protein content and changes in forager choice for abundant versus high protein flowers. Agric Ecosyst. Environ 2021;322:107645.

[23] Danner N, Molitor AM, Schiele S, Härtel S, Steffan-Dewenter I. Season and landscape composition affect pollen foraging distances and habitat use of honey bees. Ecol Appl 2016;26(6):1920–9.

[24] Pfister B, Zeeman SC. Formation of starch in plant cells. Cell Mol Life Sci 2016;73:2781–807.

[25] Parachnowitsch AL, Manson JS, Sletvold N. Evolutionary ecology of nectar. Ann Botany 2019;123(2):247–61.

[26] Stevenson PC, Nicolson SW, Wright GA. Plant secondary metabolites in nectar: impacts on pollinators and ecological functions. Func Ecol 2017;31(1):65–75.

[27] Slavković F, Dogimont C, Morin H, Boualem A, Bendahmane A. The genetic control of nectary development. Trends Plant Sci 2021;26(3):260–71.

[28] Zhao Z, Lu N, Conner JK. Adaptive pattern of nectar volume within inflorescences: bumblebee foraging behavior and pollinator-mediated natural selection. Sci Rep 2016;6(1):34499.

[29] Wright GA, Nicolson SW, Shafir S. Nutritional physiology and ecology of honey bees. Annu Rev Entomol 2018;63(1):327–44.

[30] Wright GA, Baker DD, Palmer MJ, Stabler D, Mustard JA, Power EF, et al. Caffeine in floral nectar enhances a pollinator's memory of reward. Science 2013;339(6124):1202–4.

[31] Keller A, McFrederick QS, Dharampal P, Steffan S, Danforth BN, Leonhardt SD. (More than) Hitchhikers through the network: the shared microbiome of bees and flowers. Curr Opin Insect Sci 2021;44:8–15.

[32] Steffan SA, Dharampal PS, Kueneman JG, Keller A, Argueta-Guzmán MP, McFrederick QS, et al. Microbes, the 'silent third partners' of bee–angiosperm mutualisms. Trends Ecol Evol 2024;39(1):65–77.

[33] Rering CC, Beck JJ, Hall GW, McCartney MM, Vannette RL. Nectar-inhabiting microorganisms influence nectar volatile composition and attractiveness to a generalist pollinator. New Phytol 2018;220(3):750–9.

[34] Schaeffer RN, Rering CC, Maalouf I, Beck JJ, Vannette RL. Microbial metabolites elicit distinct olfactory and gustatory preferences in bumblebees. Biol Lett 2019;15(7):20190132.

[35] Cullen NP, Fetters AM, Ashman T-L. Integrating microbes into pollination. Curr Opin Insect Sci 2021;44:48–54.

[36] Anderson KE, Sheehan TH, Mott BM, Maes P, Snyder L, Schwan MR, et al. Microbial ecology of the hive and pollination landscape: bacterial associates from floral nectar, the alimentary tract and stored food of honey bees (Apis mellifera). PLoS ONE 2013;8(12):e83125.

[37] Grienenberger E, Quilichini TD. The toughest material in the plant kingdom: An update on sporopollenin. Front Plant Sci 2021;12:703864.

[38] Marchant DB, Walbot V. Anther development—the long road to making pollen. Plant Cell 2022;34(12):4677–95.

[39] Amador GJ, Matherne M, Waller DA, Mathews M, Gorb SN, Hu DL. Honey bee hairs and pollenkitt are essential for pollen capture and removal. Bioinspir Biomim 2017;12(2):026015.

[40] Dobson HE. Survey of pollen and pollenkitt lipids—chemical cues to flower visitors? Am J Botany 1988;75(2):170–82.

[41] Roulston TH, Cane JH. Pollen nutritional content and digestibility for animals. Plant Syst Evol 2000;222(1-4):187–209.

[42] Hrassnigg N, Crailsheim K. Differences in drone and worker physiology in honeybees (Apis mellifera). Apidologie 2005;36(2):255–77.

[43] DeGrandi-Hoffman G, Chen Y, Rivera R, Carroll M, Chambers M, Hidalgo G, et al. Honey bee colonies provided with natural forage have lower pathogen loads and higher overwinter survival than those fed protein supplements. Apidologie 2016;47(2):186–96.

[44] Brys MS, Strachecka A. The key role of amino acids in pollen quality and honey bee physiology—a review. Molecules 2024;29:2605.

[45] Herbert Jr EW, Shimanuki H. Chemical composition and nutritive value of bee-collected and bee-stored pollen. Apidologie 1978;9(1):33–40.

[46] Borghi M, Perez de Souza L, Yoshida T, Fernie AR. Flowers and climate change: a metabolic perspective. New Phytol 2019;224(4):1425–41.

[47] Dai A. Increasing drought under global warming in observations and models. Nat Clim Change 2013;3(1):52–8.

[48] Pachauri RK, Allen MR, Barros VR, Broome J, Cramer W, Christ R. Climate change 2014: synthesis report. Contribution of Working Groups I, II and III to the fifth assessment report of the Intergovernmental Panel on Climate Change. Intergovernmental Panel Clim Change; 2014, pp. 151.

[49] Descamps C, Quinet M, Jacquemart AL. The effects of drought on plant–pollinator interactions: what to expect? Environ Exp Botany 2021;182:104297.

[50] Phillips BB, Shaw RF, Holland MJ, Fry EL, Bardgett RD, Bullock JM, et al. Drought reduces floral resources for pollinators. Glob Change Biol 2018;24(7):3226–35.

[51] Rering CC, Franco JG, Yeater KM, Mallinger RE. Drought stress alters floral volatiles and reduces floral rewards, pollinator activity, and seed set in a global plant. Ecosphere 2020;11(9):e03254.

[52] Descamps C, Boubnan N, Jacquemart AL, Quinet M. Growing and flowering in a changing climate: effects of higher temperatures and drought stress on the bee-pollinated species Impatiens glandulifera Royle. Plants 2021;10(5):988.

[53] Nicolson SW, Pacini E, Nepi M. Nectar production and presentation. *Nectaries and nectar*. Dordrecht, The Netherlands: Springer; 2007. p. 167–214.

[54] Lechantre A, Draux A, Hua HAB, Michez D, Damman P, Brau F. Essential role of papillae flexibility in nectar capture by bees. Proc Natl Acad Sci U S A 2021;118(19). e2025513118.

[55] Russell KA, McFrederick QS. Elevated temperature may affect nectar microbes, nectar sugars, and bumble bee foraging preference. Microb Ecol 2022;84(2):473–82.

[56] Borghi M, Fernie AR. Floral metabolism of sugars and amino acids: implications for pollinators' preferences and seed and fruit set. Plant Physiol 2017;175:1510–24.

[57] Koti S, Reddy KR, Reddy VR, Kakani VG, Zhao D. Interactive effects of carbon dioxide, temperature, and ultraviolet-B radiation on soybean (Glycine max L.) flower and pollen morphology, pollen production, germination, and tube lengths. J Exp Botany 2005;56(412):725–36.

[58] Rosenberger NM, Hemberger JA, Williams NM. Heatwaves exacerbate pollen limitation through reductions in pollen production and pollen vigour. AoB Plants 2024;16(5):plae045.

[59] Walters J, Fisher R, Sharkey TD, Isaacs R, Santiago JP. Extreme heat affects blueberry pollen nutrition, bee health, and plant reproduction. Sci Rep 2025;15(1):6249.

[60] Descamps C, Quinet M, Jacquemart A- L. Climate change-induced stress reduce quantity and alter composition of nectar and pollen from a bee-pollinated species (Borago officinalis, Boraginaceae). Front Plant Sci 2021;12:755843.

[61] LeConte Y, Spreng L, Poitout SH. Brood pheromone can modulate the feeding behavior of Apis mellifera workers (Hymenoptera: Apidae). J Econ Entomol 1995;88(4):798–804.

[62] Pankiw T. Brood pheromone regulates foraging activity of honey bees (Hymenoptera: Apidae). J Econ Entomol 2004;97(3):748–51.

[63] Forcone A, Aloisi PV, Ruppel S, Muñoz M. Botanical composition and protein content of pollen collected by Apis mellifera L. in the north-west of Santa Cruz (Argentinean Patagonia). Grana 2011;50(1):30–9.

[64]] Walters J, Barlass M, Fisher R, Isaacs R. Extreme heat exposure of host plants indirectly reduces solitary bee fecundity and survival. Proc R Soc B: Biol Sci 2024;291:20240714.

[65] Rusterholz HP, Erhardt A. Effects of elevated CO2 on flowering phenology and nectar production of nectar plants important for butterflies of calcareous grasslands. Oecologia 1998;113(3):341–9.

[66] Ziska LH, Pettis JS, Edwards J, Hancock JE, Tomecek MB, Clark A, et al. Rising atmospheric CO2 is reducing the protein concentration of a floral pollen source essential for North American bees. Proc R Soc B: Biol Sci 2016;283:20160414.

[67] Erhardt A, Rusterholz HP, Stöcklin J. Elevated carbon dioxide increases nectar production in Epilobium angustifolium L. Oecologia 2005;146(2):311–7.

[68] Marino GP, Kaiser DP, Gu L, Ricciuto DM. Reconstruction of false spring occurrences over the southeastern United States, 1901–2007: an increasing risk of spring freeze damage? Environ Res Lett 2011;6(2):024015.

[69] Inouye D. Effects of climate change on phenology, frost damage, and floral abundance of montane wildflowers. Ecology 2008;89(2):353–62.

[70] Inouye DW. Climate change and phenology. Wiley Interdiscip Rev. Clim Change 2022;13(3):e764.

Chapter 8

Climate change effects on the annual colony cycle

The annual colony cycle for honey bees is centered around brood rearing and is driven by weather and the pollen and nectar from flowering plants. From spring through the early fall, bees use the floral resources they collect to rear brood and grow the colony. As fall approaches, brood rearing slows and pollen and nectar are stored in the hive for the winter.

All parts of the annual colony cycle are connected. Factors affecting colony growth during one season will set a course for the colony to thrive or perish in future seasons. Climate change effects on temperature and precipitation impact the colony cycle through heat and cold that challenge the bees' ability to thermoregulate their hives, and through nutritional stress. Plants produce few if any flowers when under stress from heat waves and droughts. The pollen can have lower concentrations of essential nutrients, and if nectar is secreted, it may lack moisture, making it difficult for foragers to collect. Effects of nutritional stress in the spring or summer that limit brood rearing and colony growth will reverberate throughout the year especially as colonies prepare for winter.

Honey bee colony nutritional needs, physiology, population sizes, and behaviors change throughout the year in response to the seasonal temperatures, daylength, and cycles of precipitation. The types of flowering plants that are available, the nutrients they provide, and whether conditions are favorable for foraging largely depend on the season. Activities inside the hive also change with the season, most notably the production of brood. Colonies choose how to use the pollen and nectar they collect. From late winter to the fall, resources are used to rear brood. In late fall, particularly in temperate areas, nectar and pollen are stored in the bee itself, specifically in the fat body, and in the comb cells as honey and bee bread.

The colony's annual cycle begins when brood rearing resumes in mid-winter (Fig. 8.1). Brood rearing increases along with colony growth throughout the spring. The growth is fueled by large expanses of spring flowers in open fields or in dense blooms on fruit trees (Fig. 8.2). The abundant resources also fuel colony reproduction through swarming.

The Honey Bee and Climate Change. DOI: https://doi.org/10.1016/B978-0-443-28995-8.00005-2

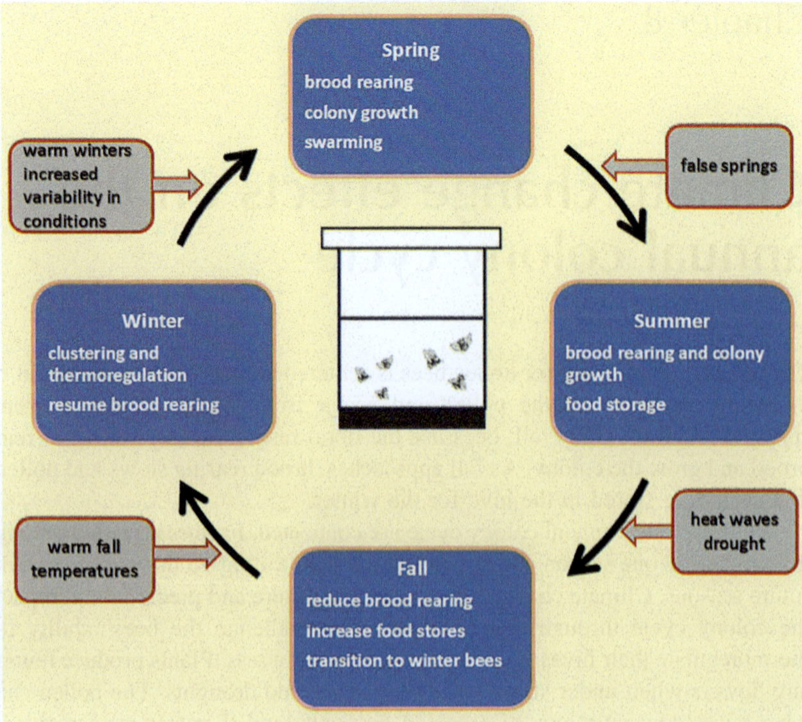

FIGURE 8.1 The annual cycle of a honey bee colony. Blue boxes describe the activities occurring during each season. The gray boxes describe possible stress factors caused by climate change. *Figure created by Gloria DeGrandi-Hoffman.*

Colonies continue to grow during the summer until August or September when brood rearing slows especially in temperate areas or high elevations and colonies begin storing resources for winter. As days become cold and short, bees form a cluster in the hive, and remain there throughout the winter. Sometime in late January or February, brood rearing resumes, and the annual cycle starts again.

As a colony moves through its annual cycle, it becomes evident that the population at any time is actually the product of past occurrences. Stress factors that affect food collection, brood rearing, or colony health during one part of the annual cycle might be felt long after the pressure is eased. For example, colony losses in the spring can have their origins during the previous summer if brood rearing is curtailed by nutritional stress or disease. Conditions in the fall that limit food storage or extend the period when colonies continue to forage after brood rearing has slowed or stopped also can set up colony losses in the spring [1]. The reckoning for high Varroa mite populations in the

FIGURE 8.2 Honey bee colonies in an almond orchard in bloom and foraging on almond blossoms. Almonds are an example of a spring bloom with numerous flowers presenting abundant nectar and pollen that colonies use to build populations in the spring.
Photograph by Emily Watkins DeJong.

fall also occurs during winter and early spring and can cause even strong colonies to fail [2].

The annual cycle in honey bee colonies supports two major goals that need to be accomplished for colonies to survive in a given year and across time. The first is building large populations in spring and early summer so that colonies can reproduce by swarming. The second is to ensure that colonies achieve sizes and store enough resources to overwinter. In this chapter, I will explain how climate change is affecting each part of the colony's annual cycle and undermining adaptive strategies to achieve both goals. For beekeepers, changes in climate are altering ways that colonies are managed and locations where apiaries can be established so colonies can thrive.

8.1 Spring

The colony cycle begins in mid-winter when brood rearing resumes often while bees still are confined in the hive. The first workers reared each year replace older overwintered bees that die shortly after foraging resumes. The new workers reared before foraging resumes are critical for colony growth, because they ease the population decline that often happens in the spring. The decline is called "spring dwindling" and occurs when older overwintered bees die faster than they are being replaced by newly emerged workers. How deeply the population declines during spring dwindling is a major determinant of how large the colony can become over the summer even if conditions are ideal for colony growth [3].

The time when bees resume rearing brood in late winter and amount of brood they rear can act as a buffer against spring dwindling. New adult workers will emerge 3 weeks after egg laying begins. If colonies start rearing brood in mid-winter (e.g., late January or early February), they can complete several brood cycles and supply the colony with new adult workers before foraging resumes in late March or April. Those colonies can have populations that are 4 times larger and swarm much earlier (i.e., mid-late May vs. June to early July) than those where brood rearing is delayed until mid-April [1]. With a healthy queen and abundant food, a positive feedback loop is established in the spring where colonies with large numbers of young bees can maintain vigorous brood rearing that creates large populations of young bees and growing colony populations going into summer.

Colony growth in the spring depends on abundant nectar and pollen. The onset of plant growth in the spring is largely driven by temperature and daylength (i.e., photoperiod). The diversity of flowering plants can be greatest in spring and supply bees with a range of pollens to meet nutritional needs [4]. In recent years, however, spring plants have begun leafing out and showing first bloom earlier in the year due to increasing global temperatures (see ref. [5]). On average, in the western United States, the time when plants begin to leaf out has been advancing for many decades [6,7]. Although periods of warm weather in late winter might seem like signs of early spring, if plants are leafing out, they could be lost by untimely frosts.

Honey bees might be especially vulnerable to the perils of early springs. Colonies begin rearing brood in late January to early February. Although warmer springs might stimulate the queen to lay eggs, bees need pollen to feed larvae and rear them into adults. If plants are not yet flowering, early season brood rearing will depend solely on the pollen remaining in the hive from the previous fall. Warmer temperatures in the late winter and early spring also will stimulate bees to forage. The physiological stress of foraging shortens the life of bees [8,9]. If the previous fall was warm and extended the period when bees could forage before they overwintered, populations could decline sharply during early spring as the older workers die at rates that exceed the emergence of new bees. The combination of earlier foraging and limited pollen for brood rearing make it difficult for colonies to slow spring dwindling and avoid the negative feedback loop that moves the population in an accelerating downward spiral.

Early springs and greater fluctuations in temperatures can limit flowering especially if early warming is followed by frosts. Seed set could also be hindered if there are inadequate pollinator populations. In temperate climates, many plants that flower in early spring rely on honey bees for pollination. Honey bees often are the most abundant pollinators in early spring and are actively foraging before other bees emerge from overwintering. However, colonies are just starting to build in early spring. When higher spring temperatures accelerate plant development and flowering, colony populations may be too small to generate large foraging populations. Fewer foragers leads to decreased flower visitation and pollination and lower seed set [10,11].

The combination of warm temperatures in late winter or early spring that cause bud break and leaf formation, only to be followed by freezing conditions are called “false springs” [5]. False springs usually occur when there are warm temperatures in March, followed by freezing conditions in April. Plant tissue including buds are tolerant of cold while in a dormant state. However, warm temperatures that stimulate plants to leaf out and progress to flowering also make them vulnerable to damage from frost (see ref. [5]). Flower structures such as petals, pistils, ovaries, and anthers are especially sensitive to frost (see ref. [12]). Damage from frost can misshape flower parts and make them less attractive to pollinators. Frost damage to reproductive parts can prevent seed set and reduce the numbers and spread of spring flowers in regions where false springs occur.

Models of plant development that incorporate future temperature regimes predict that during this century, chances of false springs will increase in some regions and decrease in others [5]. In the Great Plains and eastern forests of the United States, more frequent false springs are expected. The warm temperatures and available pollen will stimulate colonies to rear brood. However, if freezing temperatures occur, colonies could have their food sources literally disappear overnight. The loss of pollen sources and decline in brood rearing could affect colony size and health throughout the summer and fall [13].

8.2 Summer

Colony growth in summer lays the ground work for successful overwintering. Building and maintaining large colonies in the summer requires a steady supply of pollen to sustain vigorous brood rearing. The best landscapes to grow colonies and prepare them for winter are those with a continuous sequence of flowering plants throughout the summer and into the fall. Brood rearing slows in late summer and eventually stops in the fall particularly in colder areas. The size of a colony in late summer is the largest it can be going into winter.

Climate change has created stress on colonies in several ways during the summer. Increasing temperatures and more frequent heat waves make it more difficult for honey bees to maintain temperatures in the brood nest so that larvae and pupae can develop and survive. Although colonies are capable of maintaining brood nest temperatures even under extreme heat [14], especially if the colony is large, the amount of brood that can be reared is limited to the area that can be thermoregulated [15]. The problem of high temperatures can be compounded if humidity is high because these conditions make the evaporative cooling system in the hive less effective (see ref. [16]).

Colonies that are growing usually have combs that are almost entirely filled with brood. However, during heat waves, there can be wide ranges of temperatures across combs especially if the colony is small (≤10,000 bees). Outer edges of the brood area can reach temperatures of 40°C (104°F) or more, jeopardizing the development of larvae and pupae (see ref. [15]). Drone brood that usually is located on the periphery of the brood area might be especially vulnerable to high temperatures from heat waves. If the drones survive development under such high temperatures, their sperm production and viability, which are heat sensitive, can be reduced (see ref. [16]). Heat waves that affect drone fertility have long-term consequences for honey bee populations at large because low numbers of viable sperm in drones leads to poorly mated queens and small highly vulnerable colonies in the future.

High temperatures in the hive might also negatively affect adult bees. Thermal stress depletes the fat bodies in insects [17]. Fat bodies are the primary organ for the storage and release of energy needed for metabolism, immune responses, and the synthesis of detoxification enzymes [18]. If the fat body is depleted, bees will be less likely to muster immune responses to fight off pathogens or detoxify pesticides they might encounter.

During summer heat waves, colonies might become nutritionally stressed at a time when resources are needed to maintain large populations. Temperature affects many aspects of flowering, including how many flowers a plant produces, how long bloom lasts, and the quantity and quality of resources flowers provide [19–21]. High temperatures affect the amount of nectar flowers secrete and its sugar concentration. Nutrients in pollen particularly lipid concentrations also are affected by high temperatures [22,23] (see ref. [16]).

In addition to reducing of pollen and nectar production, heat waves disrupt flower development. Extreme and prolonged heat damages reproductive parts of a flower (i.e., stamens and pistils) so that even if pollination occurs, ovule fertilization and seed set can be severely limited. The effects of heat waves reverberate through plant communities, as reduced seed set makes it difficult for stands of plants to establish and expand across landscapes.

There is reason to be concerned about the effects of heat waves on plants, since the frequency of extreme high temperatures is increasing globally. Since the 1950s, the duration of heat waves and the cumulative heat they produce are increasing and the trend is accelerating [24,25]. In western Europe, the heat wave that occurred in June and July of 2019 broke records for high temperatures in France, Switzerland, Austria, Germany, the Czech Republic, Italy, and Spain. In many of the countries, June 2019 was the warmest month ever recorded [26]. Heat waves also are happening more frequently in places where they previously did not occur. A case in point is the North American Pacific Northwest heat wave of 2021 [27]. In British Columbia, temperatures reached 49.6°C (121°F), the highest ever recorded in Canada. The number of hot days and hot nights during the summer and the duration, frequency, area, and intensity of heat waves are projected to increase and accelerate in nearly all inhabited regions as climate change intensifies (see ref. [28]).

Heat waves directly impact plant mortality, fire risk, and crop failures, but they also increase the risk of extensive damage to ecosystems from events such as drought. Over the next century, drought risk and severity are predicted to increase across much of the subtropics and temperate areas as well as the Mediterranean. Soil moisture also will be affected by warmer temperatures that reduce snowfall and snowpack levels and increase water loss from evaporation [29].

During droughts, plants produce fewer and smaller flowers that emit fewer volatile attractants to pollinators [30]. If bees find the flowers, few will contain nectar and pollen. As in the case of heat waves, the lack of floral resources will cause nutritional stress that limits brood rearing, suppresses immunity, and increases likelihood of disease [31,32]. Although colonies might survive during the summer, the dry conditions and their effects on colony size and health will increase the chances of colony loss over the winter [19,21].

Drought affects colony growth and future survival, and the same can be said about the effects on flowering plants. Drought impacts plant reproduction and seed set in several ways. Less flowers are produced by plants that are water stressed, so fewer pollinators are attracted to the plants. Pollen viability and likelihood of ovule fertilization and seed set also are reduced. If less seed is set, the composition and diversity of plants at the community level will be affected over subsequent years [33] (see ref. [34]). As plant diversity declines, so does pollinator diversity [35]. The local loss of plant and pollinator species after drought periods may be most keenly felt in temperate regions where drought has been uncommon in the past but is expected to increase in frequency and severity in the future [36].

8.3 Fall

As summer turns to fall, colonies slow the pace of brood rearing and shift to storing resources. While this strategy seems like a solid way to prevent starvation during the winter, it also makes autumn the most precarious season of the year. As the days grow shorter, bees rear less brood and colonies no longer grow. If Varroa, diseases, nutritional stress or some other unfortunate event causes colonies to decline in late summer or fall, there is no chance to build and recover before winter.

Population size going into winter is important but so is the distribution of ages (i.e., age structure) of the adult worker bees in the colony [37]. The overwintering population needs to contain young, healthy, and long-lived bees to prevent a rapid decline in the population the following spring. Colonies that are infested with Varroa or show signs of disease in the fall despite the population size, can fail in the spring as more short-lived bees are lost than can be replaced with newly emerged workers.

For colonies to survive the winter, especially in temperate areas where bees can be confined in the hive for months, plants that flower in the fall need to provide abundant pollen and nectar flows. Bees need to store at least 20 kg or about 44 lbs of honey [1], with greater stores (up to 41 kg or about 90 lbs) required in areas with colder and longer winters. Diverse and plentiful pollen also is needed for rearing brood in mid-winter [38]. The brood reared while bees are still in the hive is essential for population growth in the spring, as it provides replacement workers and slows colony decline when foraging resumes.

Bees manage the pollen they collect in the fall differently than during spring and summer. Pollen is converted to brood when colonies are growing in spring and summer. In the fall, progressively less brood is reared and more pollen is stored. The first place bees store nutrients from pollen is in their bodies especially in their fat body. When the storage capacity in the bees is reached, workers convert pollen to bee bread and store it in the hive.

Reductions in brood rearing and increased nutrient storage in the bees are key indicators of the gradual conversion of summer to winter bees [39,40]. Winter bees are physiologically different than summer bees and have fat bodies with higher lipid and lower protein levels. The levels flip during overwintering when brood rearing resumes; fat body lipid concentrations decline and protein concentrations increase. The more brood that is reared while bees are confined in the hive, the greater the drop in lipid concentrations [41]. Since brood reared while bees are confined in the hive is key to colony growth in the spring, the fate of the colony in the spring is set the previous fall not only by colony age structure but also by the availability and storage of pollen from fall flowers.

Climate change is affecting overwintering survival of honey bee colonies in several ways. When falls are warm and dry, pollen and nectar can be limited because there are fewer flowers and bloom periods are shortened. Overwintering losses are higher in years when the preceding months are warmer and drier than normal for an area [19]. Warmer summers and falls also contribute to colony losses in the spring as high temperatures late in the summer can cause fall flowers to bloom earlier. Colonies that are still foraging and rearing brood in the late fall will use the resources they are collecting from fall flowers, leaving less to be stored for overwintering.

Warm falls, especially in temperate climates, may determine whether workers fully transition to winter bees. If brood rearing continues because of warm weather and bees do not store sufficient nutrients in their fat bodies, the workers might not become fully functional winter bees. A study comparing fat body composition and colony sizes before and after overwintering revealed that colonies rearing brood in November had bees with lower lipid levels in their fat bodies compared to those in colonies where brood rearing had stopped. The colonies with bees having lower fat body lipids also lost more than half of their bees during and immediately after overwintering. The colonies were overwintered in cold storage facilities that simulated winter conditions and confinement in the hive [41].

Colonies that do not store enough resources in the fall can become nutritionally stressed while overwintering. Under extreme cases, colonies can starve in a matter of hours over the winter if they are confined in the hive when resources are depleted. In cases where there are just enough resources for the colony to survive, too little brood might be reared to avoid a deep decline in the colony populations when foraging resumes in the spring.

Warm falls that extend the period when honey bees are foraging have other repercussions besides reducing food stores for the winter. Flight is physiologically stressful for insects and shortens the lifespan of honey bees [9,42]. If there are extended periods of favorable flight weather in the fall, the winter cluster could contain high numbers of physiologically older bees. Depending on when foraging resumes in the spring, the older bees could die at rates that exceed the emergence of new bees and cause populations to drop below levels where they can recover [37,43].

Warm fall temperatures can deplete food stores, reduce the longevity of workers in the winter cluster, and also make it more difficult to control Varroa. Temperatures that are amenable to foraging also enable mites to migrate into colonies on foragers [44,45]. Although mite migration could be occurring throughout spring and summer, it seems to have its greatest occurrences in the fall. As the frequency of detecting foragers with mites increases, so do the numbers of mites on adult bees and in brood cells. Levels of Deformed Wing Virus also increase with growing mite numbers just as colonies are preparing for winter [46].

Ideally, adult workers reared during the last brood cycles before winter will be the most long-live bees in the winter cluster. However, warmer fall temperatures and prolonged mite migration can compromise the health and longevity of the adults especially if they emerge from brood cells that are parasitized by multiple foundress mites (i.e., female mites). Multiple infestations occur when there are more mites than brood cells of suitable age to infest. The condition is most common in the fall because colonies are rearing less brood and Varroa populations, which might have been growing all summer, have fewer cells to infest. Adult workers from multiply-infested cells have higher virus levels and shorter lifespans than those infested by a single mite [47–49]. The increasing frequency of warm falls and the effects on Varroa populations might be contributing to high overwintering losses because colonies with high mite numbers and virus titers in the fall have little chance of surviving the winter [2]

Average fall temperatures across the United States and Europe have been increasing since 1970. The trend is expected to continue with warmer falls occurring with greater frequency [37]. The warmest fall on record in the United States and Europe occurred in 2024 (https://www.climatecentral.org/climate-matters/fall-is-getting-warmer-1). New management methods such as moving bees into indoor cold storage units or to higher elevations could stop late season flight and mite migration and perhaps increase stores of resources in the hive and in the bodies of bees. Colonies also could be moved to southern latitudes where bees are not confined in the hive for long periods and will continue to rear brood. However, colonies would need to be managed for Varroa and provided with flowering plants or supplemental feedings throughout the winter.

8.4 Winter

Perhaps nowhere is the elegance of interacting systems in a colony more on display than during the winter. While responding to the cold and living off their stores, there is coordination among thousands of workers to allocate honey and pollen for nourishment, thermoregulation so bees can remain active, and brood rearing that usually begins in late January or early February. There is little margin for error in the winter, as colonies can starve or become inactive and freeze if resources are not used judiciously.

Winter survival is the culmination of events that began the previous spring particularly those that affected colony growth [13]. Even in the absence of Varroa, colonies can be lost over the winter if they are small because smaller colonies are less effective at thermoregulation. Colonies also must have gathered enough resources to keep all the bees fed and to fuel activities such as thermoregulation and brood rearing.

Although winter can be seen as a season of reckoning for all that occurred from spring through fall, climate change has created a unique set of conditions that alone can cause colony losses. Winter temperatures are getting warmer. At higher latitudes, the climate is warming faster in winter than in summer. The warmer temperatures can cause bees to fly during afternoon hours in winter . While some flight during winter is good for colony health, as it allows bees to defecate outside the hive, frequent flight burns through stored resources that will be needed for thermoregulation and later for brood rearing. Colonies with depleted resources can be lost if weather turns cold again in late winter before flowering plants are available [19].

Although, in general, winters are becoming warmer, there is greater variability in weather conditions among years with increasing incidences of extreme winter weather. For example, in the winter of 2013–14 and again in 2018–19, cold air from the arctic flowed further south than usual [50]. Temperatures plummeted in northern latitudes and killing frosts occurred in the south where colonies actively forage and rear brood throughout the winter. Arctic air flowing far south has its origins in above normal sea surface temperatures that weaken the air flow from the jet stream [51].

Colonies are at their most vulnerable stage in the winter because they cannot accumulate more resources and rear enough brood should large numbers of workers perish. The effects of colony size and negative feedback loops can be quickly activated in winter without a chance of reversing momentum. When colonies lose workers, they also lose the ability to thermoregulate the hive. Bees on the outer edges of the cluster can become inactive and drop to the bottom of the hive and die, leaving fewer bees to thermoregulate the cluster. Extreme cold temperatures could cause losses of colonies that may have overwintered if conditions were more favorable. The possibility of extreme cold during the winter, especially in northern latitudes, underscores the importance of overwintering only large colonies with abundant honey stores.

8.5 Final thoughts

Each part of the annual cycle of a honey bee colony has its own attributes enmeshed in seasonal conditions, and all parts are interconnected so that colony success or failure often is the product of past events. Events that affect colony size, health and age structure during one season, create momentum for the population to move in either a positive or negative direction in the future. Stress from climate-related events can occur during any season and make it increasingly difficult to determine whether colonies will thrive throughout the spring and summer and build populations that can successfully overwinter.

Plants have annual cycles too. Eighty-seven percent of all flowering plants depend on pollinators [52], and their cycles are integrated into those of honey bees and other pollinators. The mutual relationship between plants and their pollinators causes them to thrive or struggle together. Climate events such as heat waves and droughts impact colony health, foraging, and flowering and can reduce pollination and the production of seed for plant regeneration. More importantly, reduced foraging can limit cross-pollination and the chances of plants producing hybrid seeds that can adapt to current and changing conditions [53].

Limited flower availability and diversity causes nutritional stress in honey bee colonies. While colonies can adjust by reducing the amount of brood they rear or alter the age when workers first forage, responses to nutritional stress can have long-term costs. Effects of stress from pathogens, pesticides, and other toxicants are exacerbated in colonies experiencing poor nutrition because physiological responses are energetically expensive for an organism. Colonies have increased susceptibility to pathogens when under nutritional stress because it reduces the bee's ability to generate immune responses (see ref. [54]). Pollen shortages also can limit the phytochemicals pollen provides that are antioxidants for disease prevention [55]. Although colonies may survive shortages in pollen, the pathogens in the colony will persist and might continue to increase in titers throughout the remainder of the year. A case in point is deformed wing virus that is transmitted by Varroa. Even though Varroa might be controlled, the virus persists and can reach its highest titers in the fall [46], with the greatest effects on colony survival in those that are nutritionally stressed [56].

The effects of pesticides also are more accentuated in colonies are experiencing nutritional stress [57]. Greater effects of pesticides occurs because the synthesis of detoxification enzymes requires nutrients bees obtain from pollen. The effects can be even greater if the nutritional stress is caused by a heat wave, since detoxification enzymes are made in the fat body that itself is negatively affected when there is heat stress [17]. With limited ability to detoxify pesticides, effects from the toxins on thermoregulation, food consumption, and flight are greater in nutritionally stressed colonies (see ref. [58]).

Climate effects on flowering plants that limit pollen availability and cause nutritional stress is particularly difficult for honey bees, as there are no suitable substitutes for pollen. Although there are many pollen supplements (diets with some natural pollen) and pollen substitutes (diets without pollen) for sale to beekeepers, the ability of bees to digest them [59] and access their protein can be limited. Feeding pollen substitutes reduces adult workers' lifespans [60] (see refs in [61]) and colony growth and survival. Virus and Nosema levels also can be higher in colonies fed pollen substitutes compared with those feeding on pollen [59,62]. However, there is good news about pollen supplements. If they are formulated with as little as 10% pollen, or if some pollen is available from flowering plants, colony growth and pathogen levels can be comparable with bees fed exclusively pollen [59,62]. These findings demonstrate the importance of having pollen sources that can flower and produce pollen even during the dire conditions of drought and heat waves. Those pollen sources might save colonies from the long-term consequences of climate stress during the summer.

Of all the effects of climate change, perhaps the most insidious in terms of colony survival is warmer fall temperatures. The strategies bees have adapted for overwintering including reduction in brood rearing, formation of winter bees, and food storage all are undermined by warm falls. Controlling Varroa also becomes more difficult with warm falls as higher temperatures extend the time when mites can migrate into colonies on foragers. Although warm falls do not occur every year, especially in temperate areas, they are predicted to occur more frequently (see ref. [37]). Depending on where colonies spent the summer, beekeepers may need to adjust their fall and overwintering management practices as methods used in the past might no longer prepare colonies to survive the upcoming winter and spring conditions.

Based on fossil records, honey bees have evolved over 60 million years. By adjusting their morphologies and behaviors at the individual and colony levels, honey bee have adapted their annual colony cycle so they can live in diverse bioclimatic conditions [63]. However, the climate is changing and putting pressure on honey bees to adapt to stressful conditions that are occurring with increasing frequency. Perhaps the most severe consequences of climate change are on managed bees as practices and apiary locations that helped colonies build and survive in the past are becoming less reliable. The evidence for this is the persistent high colony losses especially over the winter. Beekeeping began many centuries ago and provided a steady source of honey, wax, and pollination. Returning to a time when colonies thrive will require changes in land use practices, increased availability of flowering plants, and adapting beekeeping practices to the climate their colonies are experiencing.

References

[1] Seeley TD, Visscher PK. Survival of honeybees in cold climates: the critical timing of colony growth and reproduction. Ecol Entomol 1985;10(1):81–8.

[2] Genersch E, Von der Ohe W, Kaatz H, Schroeder A, Otten C, Büchler R, et al. The German bee monitoring project: a long term study to understand periodically high winter losses of honey bee colonies. Apidologie 2010;41(3):332–52.

[3] DeGrandi-Hoffman G, Roth SA, Loper GL, Erickson Jr EH. BEEPOP: a honeybee population dynamics simulation model. Ecol Model 1989;45(2):133–50.

[4] Lau P, Bryant V, Ellis JD, Huang ZY, Sullivan J, Schmehl DR, et al. Seasonal variation of pollen collected by honey bees (Apis mellifera) in developed areas across four regions in the United States. PLoS ONE 2019;14(6):0217294.

[5] Allstadt AJ, Vavrus SJ, Heglund PJ, Pidgeon AM, Thogmartin WE, Radeloff VC. Spring plant phenology and false springs in the conterminous US during the 21st century. Environ Res Lett 2015;10(10):104008.

[6] Ault T, Henebry G, de Beurs K, Schwartz M, Betancourt J, Moore D. The false spring of 2012, earliest in North American record. Eos Trans Am Geophys Union 2013;94:181–2.

[7] Ault TR, Macalady AK, Pederson GT, Betancourt JL, Schwartz MD. Northern Hemisphere modes of variability and the timing of spring in Western North America. J Clim 2011;24:4003–14.

[8] Page Jr., RE, Peng CY. Aging and development in social insects with emphasis on the honey bee, Apis mellifera L. Exp Gerontol 2001;36:695–711.

[9] Williams JB, Roberts SP, Elekonich MM. Age and natural metabolically-intensive behavior affect oxidative stress and antioxidant mechanisms. Exp Gerontol 2008;43:538–49.

[10] Hegland SJ, Nielsen A, Lázaro A, Bjerknes A-L, Totland Ø. How does climate warming affect plant-pollinator interactions? Ecol Lett 2009;12(2):184–95.

[11] Villagomez GM, Nürnberger F, Requier F, Schiele S, Steffan-Dewenter I. Effects of temperature and photoperiod on the seasonal timing of Western honey bee colonies and an early spring flowering plant. Ecol Evolut 2021;11(12):7834–49.

[12] Chamberlain CJ, Cook BI, García de Cortázar-Atauri I, Wolkovich EM. Rethinking false spring risk. Glob Change Biol 2019;25(7):2209–20.

[13] Requier F, Odoux JF, Henry M, Bretagnolle V. The carry-over effects of pollen shortage decrease the survival of honeybee colonies in farmlands. J Appl Ecol 2017;54(4):1161–70.

[14] Southwick EE, Moritz RF. Social control of air ventilation in colonies of honey bees, Apis mellifera. J Insect Physiol 1987;33(9):623–6.

[15] Chen J, Fisher II A, DeGrandi-Hoffman G, Ozturk C, Smith BH, Fewell JH, et al. Negative effects of excessive heat on colony thermoregulation and population dynamics in honeybees. Ecol Evolut Physiol 2026;99(1):1–18.

[16] Johnson MG, Glass JR, Dillon ME, Harrison JF. How will climatic warming affect insect pollinators? Adv Insect Physiol 2023;64:1–115.

[17] Klepsatel P, Gáliková M, Xu Y, Kühnlein RP. Thermal stress depletes energy reserves in drosophila. Sci Rep 2016;6(1):1–12.

[18] Arrese EL, Soulages JL. Insect fat body: energy, metabolism, and regulation. Annu Rev Entomol 2010;55:207–25.

[19] Switanek M, Crailsheim K, Truhetz H, Brodschneider R. Modelling seasonal effects of temperature and precipitation on honey bee winter mortality in a temperate climate. Sci Total Environ 2017;579:1581–7.

[20] Flores JM, Gil-Lebrero S, Gámiz V, Rodríguez MI, Ortiz MA, Quiles FJ. Effect of the climate change on honey bee colonies in a temperate Mediterranean zone assessed through remote hive weight monitoring system in conjunction with exhaustive colonies assessment. Sci Total Environ 2019;653:1111–9.

[21] Calovi M, Grozinger CM, Miller DA, Goslee SC. Summer weather conditions influence winter survival of honey bees (Apis mellifera) in the northeastern United States. Sci Rep 2021;11(1):1553.

[22] Descamps C, Boubnan N, Jacquemart AL, Quinet M. Growing and flowering in a changing climate: effects of higher temperatures and drought stress on the bee-pollinated species Impatiens glandulifera Royle. Plants 2021;10(5):988.

[23] Walters J, Fisher R, Sharkey TD, Isaacs R, Santiago JP. Extreme heat affects blueberry pollen nutrition, bee health, and plant reproduction. Sci Rep 2025;15(1):6249.

[24] Perkins-Kirkpatrick SE, Lewis SC. Increasing trends in regional heatwaves. Nature 2020;11(1):3357.

[25] Anderson S, Chartrand S. On the mean precipitation characteristics of North American heatwaves. Environ Res Water 2025;1(3):035001.

[26] Vautard R, van Aalst M, Boucher O, Drouin A, Haustein K, Kreienkamp F, et al. Human contribution to the record-breaking June and July 2019 heatwaves in Western Europe. Environ Res Lett 2020;15(9):094077.

[27] Emerton R, Brimicombe C, Magnusson L, Roberts C, Di Napoli C, Cloke HL, et al. Predicting the unprecedented: forecasting the June 2021 Pacific Northwest heatwave. Weather 2022;77(8):272–9.

[28] Domeisen DI, Eltahir EA, Fischer EM, Knutti R, Perkins-Kirkpatrick SE, Schär C, et al. Prediction and projection of heatwaves. Nat Rev Earth Environ 2023;4(1):36–50.

[29] Cook BI, Mankin JS, Anchukaitis KJ. Climate change and drought: from past to future. Curr Clim Change Rep 2018;4:164–79.

[30] Rering CC, Franco JG, Yeater KM, Mallinger RE. Drought stress alters floral volatiles and reduces floral rewards, pollinator activity, and seed set in a global plant. Ecosphere 2020;11(9):e03254.

[31] Alaux C, Ducloz F, Crauser D, Le Conte Y. Diet effects on honeybee immunocompetence. Biol Lett 2010;6(4):562–5.

[32] DeGrandi-Hoffman G, Chen Y. Nutrition, immunity and viral infections in honey bees. Curr Opin Insect Sci 2015;10:170–6.

[33] Tilman D, El Haddi A. Drought and biodiversity in grasslands. Oecologia 1992;89(2):257–64.

[34] Phillips BB, Shaw RF, Holland MJ, Fry EL, Bardgett RD, Bullock JM, et al. Drought reduces floral resources for pollinators. Glob Change Biol 2018;24(7):3226–35.

[35] Biesmeijer JC, Roberts SP, Reemer M, Ohlemuller R, Edwards M, Peeters T, et al. Parallel declines in pollinators and insect-pollinated plants in Britain and the Netherlands. Science 2006;313(5785):351–4.

[36] Dai A. Increasing drought under global warming in observations and models. Nat Clim Change 2013;3(1):52–8.

[37] Rajagopalan K, DeGrandi-Hoffman G, Pruett M, Jones VP, Corby-Harris V, Pireaud J, et al. Warmer autumns and winters could reduce honey bee overwintering survival with potential risks for pollination services. Sci Rep 2024;14(1450):5410.

[38] Mainardi G, Sponsler D, Minaud E, Vardakas F, Charistos L, Requier F, et al. Floral diversity enhances winter survival of honeybee colonies across climatic regions. J Appl Ecol 2025;62(6):1487–97.

[39] Mattila HR, Harris JL, Otis GW. Timing of production of winter bees in honey bee (Apis mellifera) colonies. Insectes Sociaux 2001;48(2):88–93.

[40] Knoll S, et al. The honey bee (Apis mellifera L., 1758) and the seasonal adaptation of productions. Highlights on summer to winter transition and back to summer metabolic activity. A review. Livestock Sci 2020;235(2019):104011. 104011.

[41] DeGrandi-Hoffman G, Corby-Harris V, Graham H, Watkins deJong E, Chambers M, Snyder L. The survival and growth of honey bee (Hymenoptera: Apidae) colonies overwintered in cold storage: the effects of time and colony location. J Econ Entomol 2023;116(4):1078–90.

[42] Margotta JW, Mancinelli GE, Benito AA, Ammons A, Roberts SP, Elekonich MM. Effects of flight on gene expression and aging in the honey bee brain and flight muscle. Insects 2013;4(1):9–30.

[43] Betti MI, Wahl LM, Zamir M. Age structure is critical to the population dynamics and survival of honeybee colonies. Royal Soc Open Sci 2016;3(11).

[44] DeGrandi-Hoffman G, Ahumada F, Graham H. Are dispersal mechanisms changing the host–parasite relationship and increasing the virulence of Varroa destructor (Mesostigmata: Varroidae) in managed honey bee (Hymenoptera: Apidae) colonies? Environ Entomol 2017;46(4):737–46.

[45] Smoliński S, Langowska A, Glazaczow A. Raised seasonal temperatures reinforce autumn Varroa destructor infestation in honey bee colonies. Sci Rep 2021;11(1):22256.

[46] DeGrandi-Hoffman G, Corby-Harris V, Chen Y, Graham H, Chambers M, Watkins deJong E, et al. Can supplementary pollen feeding reduce varroa mite and virus levels and improve honey bee colony survival? Exp Appl Acarol 2020;82(4):455–73.

[47] Floris I, Pusceddu M, Satta A. How the infestation level of Varroa destructor affects the distribution pattern of multi-infested cells in worker brood of Apis mellifera. Vet Sci 2020;7(3):136.

[48] Francis RM, Nielsen SL, Kryger P. Varroa–virus interaction in collapsing honey bee colonies. PLoS ONE 2013;8(3):e57540.

[49] Dainat B, Evans JD, Chen YP, Gauthier L, Neumann P. Dead or alive: deformed wing virus and Varroa destructor reduce the life span of winter honeybees. Appl Environ Microbiol 2012;78(4):981–7.

[50] Marshall KE, Gotthard K, Williams CM. Evolutionary impacts of winter climate change on insects. Curr Opin Insect Sci 2020;41:54–62.

[51] Palmer T. Record-breaking winters and global climate change. Science 2014;344(6186):803–4.

[52] Ollerton J, Winfree R, Tarrant S. How many flowering plants are pollinated by animals? Oikos 2011;120(3):321–6.

[53] Christmann S. Climate change enforces to look beyond the plant: the example of pollinators. Curr Opin Plant Biol 2020;56:162–7.

[54] Dolezal AG, Toth AL. Feedbacks between nutrition and disease in honey bee health. Curr Opin Insect Sci 2018;26:114–9.

[55] Liao LH, Pearlstein DJ, Wu WY, Kelley AG, Montag WM, Hsieh EM, et al. Increase in longevity and amelioration of pesticide toxicity by natural levels of dietary phytochemicals in the honey bee, Apis mellifera. PLoS ONE 2020;15(12):e0243364.
[56] Branchiccela B, Castelli L, Corona M, Díaz-Cetti S, Invernizzi C, Martínez De La Escalera G, et al. Impact of nutritional stress on the honeybee colony health. Sci Rep 2019;9(1):10156.
[57] Wahl O, Kurt U. Influence of pollen feeding and physiological condition on pesticide sensitivity of the honey bee Apis mellifera carnica. Oecologia 1983;59:106–28.
[58] Albacete S, Sancho G, Azpiazu C, Rodrigo A, Molowny-Horas R, Sgolastra F, et al. Bees exposed to climate change are more sensitive to pesticides. Glob Change Biol 2023;29(22):6248–60.
[59] Watkins de Jong E, DeGrandi-Hoffman G, Chen Y, Graham H, Ziolkowski N. Effects of diets containing different concentrations of pollen and pollen substitutes on physiology, Nosema burden, and virus titers in the honey bee (Apis mellifera L.). Apidologie 2019;50(6):845–58.
[60] Lamontagne-Drolet M, Samson-Robert O, Giovenazzo P, Fournier V. The impacts of two protein supplements on commercial honey bee (Apis mellifera L.) colonies. J Apic Res 2019;58(5):800–13.
[61] Noordyke ER, Ellis JD. Reviewing the efficacy of pollen substitutes as a management tool for improving the health and productivity of western honey bee (Apis mellifera) colonies. Front Sustain Food Syst 2021;5:772897.
[62] DeGrandi-Hoffman G, Chen Y, Rivera R, Carroll M, Chambers M, Hidalgo G, et al. Honey bee colonies provided with natural forage have lower pathogen loads and higher overwinter survival than those fed protein supplements. Apidologie 2016;47(2):186–96.
[63] Tihelka E, Cai C, Pisani D, Donoghue PC. Mitochondrial genomes illuminate the evolutionary history of the Western honey bee (Apis mellifera). Sci Rep 2020;10(1):14515.

Chapter 9

Climate change effects on commercial beekeeping

To put things in perspective, beekeepers and the colonies they manage support one-third of agricultural production worth at least US$235 billion worldwide [1,2]. Of the 115 crops that are the most important for direct human consumption, 87 depend on honey bees and other pollinators [2]. Over time, agriculture has become more dependent on pollinators. Without pollination by bees, global crop production would be reduced by 3%–8%, resulting in greater expansion of agricultural lands, a corresponding loss of unmanaged landscape and changes in the human diet [3].

During the last five decades, there has been about a 25% expansion in agricultural land globally, with most of it planted to pollinator-dependent crops [4]. Many crops that rely on honey bee pollination continue to increase in acreage, particularly oilseed, nut, and fruit crops (see ref. [5]). This trend is expected to continue as world wheat (*Triticum aestivum*), corn (*Zea maize*), and soybean (*Glycine max*) yields are predicted to decrease by 6%–16% for each degree Celsius temperature increase [6,7]. Higher CO_2 levels might offset some of the decreases in yields, but results from studies vary depending upon crops and other environmental conditions [8]. Pollinators are affected in several ways when agricultural expansion includes large acreages planted to single crops. The plantings often require high agrochemical inputs to control pests and diseases. There also is a loss of undisturbed habitats and reductions in floral resources and nesting sites (see ref. [9]). The result is lower numbers and diversity of pollinators and greater reliance of managed honey bee colonies.

High-intensity plantings of certain crops have increased over the past decades especially in fruit crops. For example, when apple orchards were planted with free-standing semidwarf trees, there were about 120–150 trees per acre (0.4 ha) that yielded about 400–1000 bushels per acre (16,800-42,000 lbs or 7620–19,051 kg). Current orchards have about 1000 trees per acre arranged in trellises (Fig. 9.1).

The high-density plantings increase the amount of fruiting wood in the orchard and can generate yields of 32,000–67,000 lbs of apples (14,515–30,391 kg) per acre (760–1600 bushels). The increases in acreages

The Honey Bee and Climate Change. DOI: https://doi.org/10.1016/B978-0-443-28995-8.00004-0

FIGURE 9.1 High-density commercial planting of apple trees on trellises. The trellis system allows more than 1000 trees to be planted per acre (0.4 ha) and generates yields up to twice as large as earlier plantings with free-standing semidwarf trees.
Photo by Gloria DeGrandi-Hoffman.

and planting densities are driven by the high market value of honey bee pollinated crops and globalization that has expanded markets. There has been more than a 300% increase in the fraction of agriculture that depends on insect pollination during the last half century. The number of managed honey bee colonies has increased globally too, although not at a rate that might avoid stressing global pollination capacity [10]. The increases in colonies varies across regions. For example, the United States and Germany have declining honey bee numbers, while colonies are increasing in China, Argentina, and Spain [9]. Furthermore, the availability of pollinators is unstable. Since 2008, yearly honey bee colony losses in the U.S. have averaged more than 40% and reached 60% in 2024–2025. Unmanaged bee populations also are declining due to loss of habitats, use of pesticides, and weather effects associated with climate change and cannot be depended upon for pollination in agricultural settings.

Honey bees and other pollinators have numerous connections to human health and well being. The human diet requires a range of fruits, vegetables, and seeds to provide essential amino and fatty acids and fiber and to maintain gut microbial diversity. Micronutrients such as vitamins A and C, calcium, fluoride, and folic acid also are supplied by bee-pollinated crops [11]. The importance of pollinators to human health is most clearly seen in areas where pollinator dependent crops are the primary dietary source of micronutrients. These regions have the greatest nutritional vulnerability tied to fluctuations in pollinator populations and the greatest incidence of micronutrient deficiency diseases [12]. Although supplements and probiotics are available for human health, the best sources of nutrients we need to optimize physiological functions such as energy production, immune responses, and cardiovascular health come from the foods we eat. Though pollinator-dependent plants contribute directly to food production, they also are used in the production of medicines, biofuels, fibers, construction materials, musical instruments, arts, crafts, and recreation activities [9].

The importance of honey bee pollination has grown in recent years because of more intense plantings of many crops. In many instances, pollination by wild pollinators living on the periphery of fields has become less reliable as populations are not large enough to impact yields especially for some early blooming crops. Numbers of feral honey bee colonies also have been seriously reduced beginning in the mid- to late-1980s by the introduction of tracheal and then Varroa mites. Some of the populations might have recovered in certain areas, but data do not exist to support this for many areas. What is agreed on, however, is that across North America there is much less pollination from unmanaged bee populations than in the past [13].

9.1 A brief history of beekeeping

Humans have known the value of honey bees and the honey they produce since the time of the ancient Egyptians (2600 BCE). There is evidence of early beekeeping by the ancient Greeks (650 BCE), who passed on the techniques to the Romans (150 BCE). As the Roman Empire expanded throughout Europe, so did the art of keeping bees. During Medieval times, beekeeping and honey bees themselves spread through much of the world (see ref. [14]). The bees were managed for honey, since sweeteners were limited and expensive in many geographic areas. The wax bees produced was used to make candles.

Honey bees first arrived in America in 1622, when they were shipped from England with the first settlers to the Colony of Virginia. Honey bees flourished in America because climate conditions were similar to areas in Europe where the bees originated. There also were large expanses of undisturbed fields and mature forests that supplied flowering plants and nesting sites for feral colonies.

Honey bee colonies spread along the eastern seaboard from New York to North Carolina and then to the west as settlements expanded. By the late 1700s and early 1800, honey bees were kept from the Atlantic to the Mississippi River. In the 1850s, colonies were shipped to California from the eastern states. Through swarming, honey bees spread to Oregon and Washington State [15].

Early beekeepers kept colonies in hives made of large hollowed logs, wooden boxes, clay pots, or straw skeps. Because early hives were limited in size and could not be expanded, colonies often swarmed in the spring. Beekeepers encouraged swarming as it was a means of obtaining new colonies. Honey was collected from colonies by cutting out the honey comb. If the colony survived, the bees would have to build new comb before they could expand.

Beekeeping changed dramatically in the 1850s with the invention of the movable frame. A Congregational minister from Pennsylvania, Rev. L.L. Langstroth, patented a hive where wax combs were built by bees within seperate removable frames in wooden boxes. The key to the removable frame was that they were built so that each frame was separated from the next by three-eighths of an inch. If frames were separated by less space, the bees would fill the gaps with propolis and wax. If there was more than three-eighths of an inch, the bees would fill the space with wax comb.

The removeable frame opened the door for numerous innovations and inventions that began modern beekeeping (remember, this was the age of the First Industrial Revolution). These included wax foundation for the frames that bees would use to build comb. The centrifuge extractor was invented so that honey could be nondestructively removed from the comb. The empty combs were placed back into the hive boxes for the bees to use again. With the new inventions, beekeeping and honey production became a popular and profitable enterprise. Honey bee queens were imported from Italy in the 1860s as these bees were known to build large populations that were easy to work because of their gentleness. The Italian bee colonies also produced large amounts of honey. Today, the Italian honey bee is still the predominant race of bees used in the United States (for a more in-depth discussion of the history of beekeeping, see refs [15,16]).

Moving colonies to follow blooms of flowering plants and to pollinate crops was limited by horse-drawn wagons and the lack of roadways suitable for moving bees. Attempts to move colonies by ship or rail had only marginal success. As roads improved largely because of increased adoption of travel by automobile, moving hives of bees became more common. The practice of moving hives for honey crops and for pollination has grown. Today, more than 1.5 million colonies (one-third of all colonies in the United States) are moved to California each winter for almond pollination [17]. After almonds, beekeepers move hives from farm to farm pollinating fruit, vegetable, and seed crops, charging pollination fees as the crops bloom. In recent years, U.S. beekeepers received approximately US$350 million in pollination fees [13]. Beekeepers also can move their colonies in the late spring to the Northern Great Plains for honey production. More than a million colonies are moved to this region and remain there throughout the summer and early fall.

My own personal history of beekeeping began when I was in graduate school in the 1980s at the Michigan State University. I learned beekeeping from Dr. Roger Hoopingarner who was my Ph.D advisor. The apiary was off campus in a forested area surrounded by the University's research farms. Michigan has a temperate climate with cold and frequent subfreezing temperatures from November to March. Snow fall could be heavy because of the proximity to the Great Lakes.

Although colonies could be under the snow, in late January–early February, the queen began laying eggs and colonies were transitioning from overwintering to their spring build up. By April, afternoons were getting warmer and foraging bees collected pollen and nectar from early spring flowers and blooming hardwood trees. Colonies grew rapidly as more flowers bloomed and the brood nest expanded. By May or early June, we divided our strongest colonies to prevent swarming.

Summer brought on a sequence of wildflowers that provided the bees with pollen and nectar to sustain and grow the populations built in the spring. In good years, we would continue splitting colonies and harvesting honey so colonies did not get too large and unwieldy. Later in the summer as daylength shortened, brood rearing slowed. Bees collected pollen and nectar from early fall flowers and stored them in the hive for winter. By late October, colonies stopped rearing brood and spent nights and much of the days in cluster. From November to March, there may have been a few days when it was warm enough for bees to fly and defecate (cleansing flights), but the bees mostly stayed in cluster in the hive. Sometime in late January or early February, brood rearing resumed, and the yearly colony cycle began again.

The summary of my first experiences in beekeeping sound idyllic, and, in someways compared to now, they were. Brood diseases like European and American foulbrood [18,19] occurred mainly in the spring, but often were self-limiting and cleared up with a good nectar flow. If symptoms of infections persisted, antibiotics were applied. Chalkbrood disease caused by *Aspergillus flavis* also was a problem especially in the spring and early summer when the weather could be wet and cold. Again, strong colonies could clean out diseased cells, and over time clear up the infection. Wax worms (*Galleria mellonella*) were a problem primarily in weak colonies. Combs with wax worm larvae were removed and weak colonies were combined with a stronger ones. Overwintering colonies that were confined in the hive for months could have Nosema (*Nosema apis*) infections so fumagillin would be fed to bees in sugar syrup in the fall [20]. The bees would consume and store the sugar syrup in the hive so that treatment would last over the winter. Overwintering losses were usually less than 10% for experienced beekeepers.

When I learned beekeeping, I was learning a system predicated on seasonal patterns and tried-and-true management techniques used for more than a century. Many of those techniques are still being used, but yearly colony losses are averaging 40%–60% especially for commercial colonies. Yearly losses of

40% by large-scale beekeepers are being seen as the "new normal" [21]. Many of the losses are from Varroa mites and the viruses they carry [22]. Poor nutrition and pesticide exposure also have become increasingly problematic and add to the difficulties in keeping colonies alive [23]. Land use and diminishing acreage of natural forage contribute to poor nutrition and makes colonies more vulnerable to the stresses they encounter.

Although colony losses are explained by parasites and pathogens, pesticides, and poor nutrition [23], an overarching cause is the direct and indirect effects of climate change. In this chapter, I will describe how colony losses can occur because of where bees are kept and the likelihood of weather events such as floods, droughts, and fire (direct effects). I also will discuss indirect effects of climate change and how they exacerbate the various stress factors colonies experience during each season of their yearly cycle.

9.2 Direct effects of climate change on colony survival

Extreme weather events that can cause catastrophic damage and eliminate entire apiaries are occurring more frequently over wider geographic areas [24]. The latest reports from the United Nations Intergovernmental Panel on Climate Change (IPCC) (https://www.ipcc.ch/report/ar6/wg2/) and from research studies (see refs in [25]) conclude that climate change will cause rising temperatures in summer that will persist into the fall, resulting in shifting patterns and seasons for precipitation events. The occurrence of severe and extreme weather events such as hurricanes, tornados, droughts, fire, and floods will increase. Flooding from heavy rainfall and rising sea levels as well as larger and more uncontrollable wildfires already are occurring across the globe and also will increase in frequency [25,26]. These weather events are disrupting ecosystems and plant–pollinator relationships and destroying apiaries located in their path.

Much of the severe weather that is occurring is due to rising ocean temperatures. June to November is "Hurricane season." Hurricanes (tropical cyclones) form from the warm ocean waters in the tropics or from the jet stream. Storms also can occur when cold dry air masses from northern latitudes meet warm moist air from southern regions (extratropical cyclones) (https://www.epa.gov/climate-indicators/climate-change-indicators-tropical-cyclone-activity). When the storms make landfall, intense rains, high winds, and storm surges off the ocean can cause severe flooding, soil erosion, property and infrastructure damage, and loss of life.

Apiaries along the eastern seaboard of the United States, the Caribbean and Gulf of Mexico are particularly susceptible to winds and flooding from hurricanes. In 2022, the 150 mph (241.4 km/h) winds and 13 ft (4 m) storm surge from Hurricane Ian caused an estimated 150,000–300,000 hives to be lost in Florida (https://www.nytimes.com/2022/12/14/us/hurricane-ian-florida-bees.html). In the aftermath of hurricanes Helene and Milton in September and

October of 2024, an estimated 95,000 hives were destroyed across Florida, Georgia, North Carolina, and surrounding states. Colonies that survived were nutritionally stressed as flowering plants also were destroyed from the flooding (https://greatergood.org/blog/supporting-beekeepers-after-hurricanes-helene-and-milton).

While some areas are receiving too much rain, others are experiencing too little precipitation. Dry conditions make regions susceptible to fire. Fires are part of natural ecosystems and can modify the diversity, structure, and interactions among species in ecosystems. Although fires may limit the availability of flowering plants in the short run, over time the diversity and density of flowering plants and their pollinators can increase as more open space and soil resources become available [27].

The benefits of fire are tied to their frequency. Fires can be beneficial if they occur intermittently. When there are long intervals between fires, seed bedded during previous years can germinate and plants can reestablish and spread across landscapes. However, when fires are frequent, there might not be sufficient time for plants, pollinators, and other animals to reestablish [24,28]. Recovery also may be difficult if wildfires develop into high-severity fires, where there are extensive losses of mature trees and widespread soil exposure and damage. Landscapes will take many years to recover from high-severity fires because of increased soil erosion, runoff [29–31], and disruption of soil biochemistry [32]. High-severity fire also releases large amounts of carbon into the atmosphere from burning mature trees. The fires create a positive feedback loop that increases atmospheric CO_2 concentrations causing additional climate warming that set up conditions for fires in the future [33].

In western U.S. forests, wildfire activity underwent an abrupt and sustained regional increase in the mid-1980s. Nearly two-thirds of the increase in fires were concentrated in forests of the Northern U.S. Rocky Mountains (between 428 and 508 N latitude) [34]. The change is linked to climatic factors such as higher temperatures, drier summers, below-average winter precipitation, and earlier spring snowmelt [35]. Southeastern Australia also have seen changes in the frequency and severity of wildfires. The largest and most severe fires on record occurred during the 2019/2020 fire season [36,37]. The fires were preceded by drought conditions. Fires in Canada also are getting larger (>200 ha), and fire seasons are getting longer [38].

Beekeeping is directly affected by the increasing incidence of wildfires. Apiaries in the path of the fires are lost. Colonies that survive can be weakened because flowering plants are lost especially when fires occur in grasslands. Even apiaries outside the burn areas are affected because of the smoke from the fires that can last for weeks. Smoke interferes with the bees' navigation system that relies on polarized light, so bees may stop foraging until the air is clear. If foragers leave the hive, flowers may be more difficult to find especially if odor cues are masked by smoke and ash (https://extension.oregonstate.edu/forests/fire/how-do-wildfires-affect-bees).

The severity of fire weather (hot, dry conditions) has been increasing over the last 50 years [39–41]. We now speak of "megafires," where more than 10,000 ha (27,710 acres) burn in a single fire event [27,42]. Megafires have occurred across the globe in regions such as the California, Canada, Eastern Australia, the Amazon, and the Mediterranean and Congo basins. The 2019–2020 fire in Australia was a megafire that burnt 24 million acres. In Australia's biggest beekeeping state, New South Wales (NSW), 13 million acres, 7% of the state's area, including 37% of NSW National Parks and 50% of State forests burned. The NSW Bushfires burned over 1000 hives and destroyed the foraging area for over 88,000 colonies (https://ucanr.edu/blog/bug-squad/article/bee-people-bees-and-fires).

The direct effects of climate change on honey bees have immediate impacts on apiaries and livelihoods of beekeepers in the wake of severe storms and fires. However, there also are long-term consequences. When landscapes are severely altered by climate events, forage for bees might not reestablish for many years or at all if unfavorable conditions such as hot dry summers persist. Beekeepers need undisturbed ecosystems with flowering plants for their colonies to build and thrive. The long-term effects of extreme weather events along with changes in land-use patterns and expansion of agricultural plantings will make suitable apiary sites more difficult to find especially in regions prone to extreme weather events.

9.3 Indirect effects of climate change on beekeeping

Unlike colony losses from catastrophic weather events, the indirect effects of climate change on colonies losses are subtle and difficult to track. Colony losses can occur months after climate events occur. Factors that slow or stop brood rearing in spring or summer or exert nutritional stress that shortens the lifespan of adult workers reduce colony sizes and, more importantly, their growth trajectory during spring and summer. For instance, events such as heavy rainfall interrupt foraging and can deplete colony resources. Nectar is diluted after heavy rains, and pollen might not be released from anthers for days [43]. In the colony, lack of incoming resources affects developing brood and may cause the queen to stop laying eggs. The break in brood rearing stops colony growth. If the population does not recover before fall, the colony can be lost over the winter especially if it already is weakened by other stress factors such as mites or pesticides.

Unseasonably warm temperatures in the spring can affect plant–pollinator relationships because higher temperatures cause plants to bloom earlier in the season. High temperatures also can shorten the bloom period for some plants [44]. Nectar and pollen from spring flowers are essential for colony growth after overwintering.

In the summer, colonies can be stressed by prolonged periods of high temperature, referred to as heat waves. The World Meteorological

Organization defines heat waves as periods where the maximum and minimum temperatures are unusually high for a location. The minimum temperature is as important as the maximum temperature, especially for honey bee colonies. Minimum temperatures often occur at night and can be as high as 37.8°C (100°F) when daytime temperatures are over 43.3°C (110°F). A positive feedback loop can establish because if temperatures remain warm at night, higher temperatures are reached earlier in the day and last for longer periods (https://wmo.int/topics/heatwave).

Across the United States, heat waves have gone from an average of two per year in the 1960s to more than six in the 2020s. The intensity, duration, and time during the year when they can occur also have increased since the 1960s (https://www.c2es.org/content/heat-waves-and-climate-change/). The trend is occurring across the globe at faster rates than climate models predicted [45].

Honey bees and other social insects have more resilience than solitary insects in that collectively they can regulate temperatures in the hive. However, there are limits to honey bees' ability to cool the hive and also costs in terms of colony growth. Honey bees concentrate their temperature regulation primarily in the brood area where temperatures are maintained at 34.5°C ± 1.5°C (see ref. [46]). During periods when ambient temperatures are high, workers cool the nest by fanning with their wings to increase air circulation. The bees also will collect water and spread it over the combs for evaporative cooling. Temperatures in the hive also can be moderated by workers collecting at the entrance of the hive and fanning to circulate air. If colonies have access to water, temperatures especially in and around the brood area can be maintained.

Under conditions of extreme heat, the thermoregulations system can break down. If conditions are hot and dry, bees may not be able to collect enough water to regulate temperatures. Extremely high temperatures during the day also can cause nighttime temperatures to remain high. Since bees do not forage at night, the lack of water can cause colonies to overheat. In these cases, some colonies lose adult bees and brood and are severely weakened, while others die overnight (GD-H personal observation).

When the thermoregulation system is compromised, the effects are more subtle. In Phoenix, Arizona, USA, temperatures in the summer can exceed 40°C in the shade. A study conducted there showed that colony populations declined during summer heat waves [47]. The effectiveness of the thermoregulations system was compromised during periods of extreme heat in the afternoons. Temperature ranges on the outer combs of the brood nest exceeded those required for rearing healthy brood. Smaller colonies were less effective than larger ones at cooling the hive especially over expanded areas of the brood nest.

Negative feedback loops can establish during heat waves especially in smaller colonies because of the relationship between the area that can be thermoregulated and the amount of brood that can be reared. If brood rearing is limited to areas that can be cooled and those areas get smaller as adult populations decline, progressively less brood will be reared, leading to even smaller areas that can be thermoregulated and the downward spiral begins. Although some colonies may survive, if they remain small throughout the summer, they will not grow sufficiently by the fall to survive the winter.

In addition to effects of high temperatures on the amount of brood colonies can rear due to the limits of thermoregulation, there also can be heat-induced reductions on the ability of queens to lay fertilized eggs [48]. While queens are reported to have high tolerance to heat, the sperm they store in their spermatheca after mating can lose viability. High temperatures also affect the viability of sperm in drones and the viability of drones themselves [49].

Warmer summers have led to higher temperatures in the fall. In the contiguous United States, fall temperatures have increased by 2.4°F since 1970. The Southwest (+3.8°F), northern Rockies, and Great Plains (3.1°F) have experienced the greatest temperature increases since 1970 (https://www.climatecentral.org/climate-matters/2024-fall-package). Warmer falls increase the impact of Varroa on honey bee colonies in two ways. First, warmer fall temperatures can extend the period when colonies rear brood and Varroa reproduce [50]. Warmer temperatures also extend the period when bees can forage and mites can enter colonies on foragers [51]. High mite numbers and virus titers that often go together lead to colony loss over the winter.

The second way that warm falls jeopardize overwintering survival is by extending the period when bees can fly. Ideally, bees that comprise the winter cluster are those that emerged in the fall and flew very little because of colder temperatures. Limiting flight in the fall is important to the age structure of the winter cluster. Bees physiologically age when they fly [52,53], so having cooler falls that limit flight creates a winter cluster with greater numbers of younger bees than if temperatures were warm enough for flight to occur. Winter clusters with younger bees have less population decline in the spring when foraging resumes and grow into larger colonies in the summer [54].

Although stress from climate change can be detrimental to colony nutrition and growth, Varroa mite infestations can be fatal. The arrival of Varroa has caused major shifts in how colonies are managed, including applications of miticides into colonies, effects on honey production, and overwintering survival. Interactions between Varroa and honey bee colonies are highly dynamic as the mite acquires resistance to miticides and spreads those capabilities to other colonies when they migrate into hives on foragers. Climate change and the warmer fall temperatures can extend the period of mite migration and reduce overwintering survival of infested colonies.

Whether it is colony nutrition, Varroa control, overwintering, or just choosing an apiary site, climate change is affecting the outcome of beekeeper's decisions. This new reality makes the tried-and-true colony management methods of the past less reliable. Developing and adopting new strategies and practices for year-around colony management will be essential for maintaining the solvency of commercial beekeeping.

9.4 Pesticides and climate change

Since the 1940s, there has been increased crop losses from pests on a global scale and a 33-fold increase in the amount and toxicity of pesticides used (see ref. [55]). Many of those pesticides have found their way into honey bee colonies. A colony of honey bees will forage for up to 2 miles from the hive and pick up pesticides along with the pollen and nectar they collect. The nectar and pollen are stored in wax comb. Many pesticides are wax soluble so they can penetrate the waxy cuticle of insects. If pollen and nectar are contaminated with wax-soluble pesticides and are stored in comb cells, the pesticides will leech out and bind to the wax. Since wax comb is infrequently replaced by beekeepers, the pesticides can accumulate for several years. The result is that the wax comb that bees use to rear brood and store food can be highly contaminated with pesticides.

Surveys of honey bee colonies have found pesticide contamination in hives throughout the world (see ref. [56]). In the United States, a survey of wax comb samples detected 87 different pesticides. Miticides to control Varroa were the most commonly found in the wax comb, but the remainder were agricultural pesticides detected in a majority of samples. On average, eight different pesticide residues were found in each sample, but up to 39 different pesticides could occur. In 350 samples of incoming pollen, 98 pesticides were identified, with up to 31 different pesticides found in a single sample. In addition to insecticides, high levels of fungicides were commonly found in pollen samples [57]. Similar results were found in a second study with samples of wax comb and pollen collected by foraging bees in areas and states not previously sampled in the earlier study. Depending on the location, one to nine compounds were detected in the pollen samples, including insecticides, fungicides, and herbicides. Twenty-nine different fungicides were found, with most occurring in pollen samples. Again, miticides for Varroa control were found in most wax comb [58].

In Italy, a 3-year survey of pollen collected from foragers as they entered the hive showed that a majority of the pollen samples (62%) had at least one pesticide and more often were contaminated with multiple compounds. The most commonly found insecticide was chlorpyrifos [59]. In Spain, a study that included analysis of bees inside the hive, stored pollen, and wax comb revealed that pollen, especially if it was stored in comb cells, contained the largest number of pesticides. Miticides used for Varroa control were found, but pollen also had insecticides applied to agricultural crops. Concentrations of the insecticides were highest in apiaries located in intensive farming areas [56].

The persistent exposure to sublethal pesticide contamination has long-term effects on honey bee colony survival. In addition to acute exposure causing colony death, stress caused by sublethal pesticide exposure during honey bee development and as adults can cause extended periods of immunosuppression, opening the way to infections by viruses and other pathogens. Foraging, sperm viability, and gene expression also can be affected (see ref. [56]). Sublethal chlorpyrifos exposure can reduce queen emergence, possibly due to the suppressed immunity and high virus levels found in developing queens [60]. Consuming pollen with high fungicide residues also has sublethal effects on honey bees. Fungicides can compromise respiration and reduce the bees' ability to digest protein, causing nutritional stress and high virus titers [61]. Consuming pollen with fungicides also increased bees' likelihood of being infected with the gut parasite *Nosema cerana* [62].

The widespread use of a class of insecticides called neonicotinoids also is a great concern for the health of honey bees and other pollinators, as well as beneficial insects in general. Neonicotinoids are neurotoxins and are highly effective in controlling a wide range of pest insects. They are systemic insecticides, meaning that they are absorbed by the roots and leaves of plants and distributed among plant tissues. When insects consume leaves or pierce and suck the fluids from plants, they ingest fatal amounts of toxic neonicotinoids [63]. In crops such as corn and soybean, neonicotinoids are applied directly to the seed as a powder so that it is absorbed by the roots as the seed germinates and is transferred to plant tissue as the plant grows. In citrus orchards, neonicotinoids are added to irrigation water and are taken up by the tree roots [64]. In other crops, neonicotinoids are applied as foliar sprays. Applied more often than any other class of insecticides, neonicotinoids are licensed for use in more than 120 countries and comprise approximately one-quarter of all insecticides used [63].

Neonicotinoids can persist and accumulate in soils and because they are water soluble and can leach into waterways and field margins. The insecticide has been detected in noncrop plants growing in field margins and soils in organic farms and undisturbed ecosystems [63,65]. Neonicotinoids also have been detected in pollen and nectar of flowering plants adjacent to agricultural fields planted with neonicotinoid-treated seeds [63]. In some instances, honey bees encounter sublethal levels of neonicotinoids in nectar and pollen. A study showed that honey bee colonies located near corn crops treated with neonicotinoids have increased worker mortality, reduced social immunity, and increased queen loss over time. The toxicity of neonicotinoids to honey bees doubles in the presence of certain fungicides [66]. Queen rearing may also be affected if pollen is contaminated with neonicotinoids. A study where queen rearing colonies were fed pollen supplements containing field realistic levels of neonicotinoids found that the queens had abnormal ovaries and reduced amounts of stored spermatozoa within their spermathecae. During the 4-week study period, fewer queens exposed to neonicotinoids during development produced worker brood and survived compared with queens reared in colonies without exposure to the compounds [67].

Colony losses from lethal and sublethal pesticide exposure are likely to increase due to climate change as pest problems become more frequent and severe. Increases in pesticides use could occur for several reasons. The persistence and efficiency of pesticides are impacted by the intensity and timing of rainfall and by temperature. Pesticide chemistries are heat sensitive and high temperatures can reduce their effectiveness in controlling pests. High temperatures also influence insect metabolism and increase enzymatic efficiency and detoxification of compounds [68]. What seems like resistance to a pesticide might be due to high ambient temperatures. Under cooler conditions, the insect might be susceptible to the toxic effects.

Temperature and precipitation are major factors controlling the population growth, development, and spatial and temporal distribution of insects, weeds, and pathogens [55]. Insect pest populations could become more abundant because warmer temperatures raiseoverwintering survival, enhance dispersal, accelerate reproduction and development, and potentially increase the number of generations insects can complete per year [55,69,70]. With higher temperatures, insect may become active earlier and increase pressure on crops. For example, surveys encompassing nearly 50 years of sampling in the United Kingdom show that 55 aphid species, including most key pests of grains and other crops, currently have earlier first flights than in the past [71]. Aphids pierce plant tissue and feed on the sugary fluid (sap) circulating through the phloem. While feeding, aphids can transmit viruses. Earlier first flights causes pest pressure to occur earlier in the season and this increases both direct injury and virus levels in plants [72].

As areas become warmer, pests may expand their geographical range, and this could impact the effectiveness of biological control programs for insect pests. Phenological overlap between insect pests and their predators and parasites can be disrupted by range shifts and cause pest species to escape from their natural enemies. Range shifts also can cause predators to find alternative food sources and lose their impact as biocontrol agents [73] (see ref. [70]).

In recent decades, globalization has led to an increase in the worldwide movement of people and goods and an increase in the number of insects introduced into areas outside their original range. The harmful effects of invasive insects may be expanded because of climate change as barriers to the successful establishment and dispersal of certain species decrease [74]. Plant pathogens also could spread leading to increased use of fungicides especially since increased rainfall produces wet conditions that are ideal for fungal spore germination and spread (see ref. [69]). Further compounding the impact of pests and pathogens is that exposing plants to high heat or dry conditions reduces their natural defenses, especially in the early stages of growth, making plants more susceptible to insect feeding and infections by pathogens.

Currently, insect pests account for 10%–28% of agricultural yield losses [75]. Greater losses could occur in the future because of increases in pest abundance and climate change-driven reductions in crop performance and yield. If pest problems are solved only by greater use of pesticides, the environmental problems (i.e., loss of biodiversity, reductions in pollinator populations, etc.) that are occurring today will increase. A commitment to preserving pollinators in particular and biodiversity in general requires the development of new systems to produce crops and maintain an affordable food supply. Modifications to farming systems designed to mitigate and adapt to climate change can help preserve pollinators and biocontrol agents while controlling pest species. Some practices to reduce pesticide use could include reduced tillage (i.e., turning over the soil to prepare seed beds or remove crop residue), increased crop diversification, intercropping, and cover crops. These practices can slow erosion, improve soil health and water availability, reduce weed species, fix nitrogen, enhance nutrient cycling efficiency, improve carbon capture, increase crop productivity, and stabilize crop yields [70]. There also is some evidence that these practices can increase natural enemies and reduce pest populations, but the results are not consistent, indicating that we still have much to learn (see ref. [70]). Pest pressure is predicted to increase through the 21st century, but the answer cannot be to increase pesticide use. Anticipating and responding to the effects of climate change by designing systems that stabilize and even enhance crop production while preserving the environment are essential and will require interdisciplinary efforts to understand the complex interactions among soils, plants, pathogens, and insects in agricultural systems.

9.5 Future directions

Systems that persist over time adapt by disassembling subsystems that no longer are useful and constructing new ones in response to current conditions. This is easier said than done in beekeeping operations. Commercial beekeeping businesses often are family owned and have existed for generations. Investments, both monetary and emotional, have shaped where the business is today. Without adapting to current conditions, however, colony losses will remain high, putting the future of many beekeeping businesses in peril.

The repercussions of multiple stress factors on the survival of honey bee colonies and solvency of beekeeping operations are already being realized. In the United States, the market size of the beekeeping industry declined by 2.9% between 2020 and 2025, and further reductions are forecasted for at least the next 5 years (https://www.ibisworld.com/united-states/industry/beekeeping/68/#TopQuestionsAnswered). The population declines of honey bees and other pollinators have affected yields in crops such as cherries and blueberries [76]. Oilseed crops that benefit from foraging honey bees and other pollinators could also be affected [77,78].

The best way to develop new subsystems to reduce colony losses is to clearly consider the current environment and prepare for future conditions. Controlling Varroa population growth without increasing the use of miticides needs to be the top priority. In addition to mite resistance, the accumulation of miticides in the wax comb negatively impacts queen and drone reproductive health, brood development, worker bee learning, longevity, and colony strength [79,80]. Control strategies that incorporate mite resistant lines of honey bees and miticides that do not accumulate in wax comb are control measure that are currently available. Control strategies specific to colonies located in different geographic regions and managed for pollination, honey production, and queen rearing also need to be developed.

Beekeepers cannot solve the problem of colony losses alone. Poor nutrition creates a major stress on colony growth in the spring and summer and on preparation for overwintering in the fall. Areas where colonies are kept now might not be suitable in the future due to the likelihood of climate incidents such as fire, flood, heat waves, and drought. New apiary sites will be needed, and areas with abundant forage should be made available to beekeepers. Safe spaces that reduce pesticide exposure could improve the health of queens and reduce colony losses from the effects that sublethal pesticide exposure has on immunity, longevity, and behavior. Systems developed to keep colonies alive and promote a vibrant beekeeping industry also should include subsystems for overwintering management. These could include preparing areas with flowering plants for overwintering in southern latitudes and recommendations for managing colonies that will overwinter in cold storage.

Honey bees once thrived in the areas where annual colony losses now can exceed 40%. What has changed is the landscapes where apiaries are located, the pesticide levels bees are encountering, and the introduction of Varroa mites. Climate change will continue to make beekeeping more difficult and jeopardize the future of this industry. We all stand to lose if beekeeping is not protected. Our health depends on the crops where honey bees are needed for pollination. Discussions about changing agricultural practices that incorporate effects of climate change need to include methods that preserve and protect honey bees and other pollinators and value these natural resources as much as stable crop yields.

References

[1] Gallai N, Salles JM, Settele J, Vaissière BE. Economic valuation of the vulnerability of world agriculture confronted with pollinator decline. Ecol Econ 2009;68(3):810–21.

[2] Klein AM, Vaissière BE, Cane JH, Steffan-Dewenter I, Cunningham SA, Kremen C, et al. Importance of pollinators in changing landscapes for world crops. Proc R Soc B 2007;274(1608):303–13.

[3] Aizen MA, Garibaldi LA, Cunningham SA, Klein AM. How much does agriculture depend on pollinators? Lessons from long-term trends in crop production. Ann Bot 2009;103(9):1579–88.

[4] Aizen MA, Garibaldi LA, Cunningham SA, Klein AM. Long-term global trends in crop yield and production reveal no current pollination shortage but increasing pollinator dependency. Curr Biol 2008;18(20):1572–5.

[5] Aizen MA, Aguiar S, Biesmeijer JC, Garibaldi LA, Inouye DW, Jung C, et al. Global agricultural productivity is threatened by increasing pollinator dependence without a parallel increase in crop diversification. Global Change Biol 2019;25(10):3516–27.

[6] Asseng S, Ewert F, Martre P, Rötter RP, Lobell DB, Cammarano D, et al. Rising temperatures reduce global wheat production. Nature Clim Change 2015;5(2):143–7.
[7] Kucharik CJ, Serbin SP. Impacts of recent climate change on Wisconsin corn and soybean yield trends. Environ Res Lett 2008;3(3):034003.
[8] Kimball BA. Crop responses to elevated CO2 and interactions with H2O, N, and temperature. Curr Opin Plant Biol 2016;31:36–43.
[9] Potts SG, Imperatriz-Fonseca V, Ngo HT, Aizen MA, Biesmeijer JC, Breeze TD, et al. Safeguarding pollinators and their values to human well-being. Nature 2016;540(7632):220–9.
[10] Aizen MA, Harder LD. The global stock of domesticated honey bees is growing slower than agricultural demand for pollination. Curr Biol 2009;19(11):915–8.
[11] Smith MR, Singh GM, Mozaffarian D, Myers SS. Effects of decreases of animal pollinators on human nutrition and global health: a modelling analysis. Lancet 2015;386(10007):1964–72.
[12] Chaplin-Kramer R, Dombeck E, Gerber J, Knuth KA, Mueller ND, Mueller M, et al. Global malnutrition overlaps with pollinator-dependent micronutrient production. Proc Biol Sci 2014;281(1794):20141799.
[13] Rucker RR, Thurman WN, Burgett M. Honey bee pollination markets and the internalization of reciprocal benefits. Am J Agric Econ 2012;94(4):956–77.
[14] vanEngelsdorp D, Meixner MD. A historical review of managed honey bee populations in Europe and the United States and the factors that may affect them. J Invert Pathol 2010;103:S80–95.
[15] Oertel E. History of beekeeping in the United States. In: Martin EC, Oertel E, Nye NP, editors. Beekeeping in the United States. Agriculture Handbook No. 335 (Revised). U.S. Department of Agriculture; 1980. p. 2–9.
[16] Mangum WA. A brief history of beekeeping n the United States. In: G JM, editor. The Hive and the Honey Bee. Hamilton, IL, USA: Dadant and Sons Inc.; 2015. p. 25–50.
[17] Bond JK, Hitaj C, Smith D, Hunt K, Perez A, Ferreira G. Honey Bees on the Move: From Pollination to Honey Production and Back. Economic Research Report No. 290. U.S. Department of Agriculture, Economic Research Service; 2021. p. 1–27.
[18] Ratnieks FLW. American foulbrood: the spread and control of an important disease of the honey bee. Bee World 1992;73(4):177–91.
[19] Fosgren E. European foulbrood in honey bees. J Invert Pathol 2010;103:S5–9. 2010.
[20] Peirson M, Pernal SF. A systematic review of Fumagillin field trials for the treatment of Nosema disease in honeybee colonies. Insects 2024;15(1):29.
[21] Cilia L. The plight of the honeybee: a socioecological analysis of large-scale beekeeping in the United States. Sociol Ruralis 2019;59(4):831–49.
[22] Genersch E, Von der Ohe W, Kaatz H, Schroeder A, Otten C, Büchler R, et al. The German bee monitoring project: a long term study to understand periodically high winter losses of honey bee colonies. Apidologie 2010;41(3):332–52.
[23] Hristov P, Shumkova R, Palova N, Neov B. Factors associated with honey bee colony losses: a mini-review. Vet Sci 2020;7(4):166.
[24] Neumann P, Straub L. Beekeeping under climate change. J Apic Res 2023;62(5):963–8.
[25] Insolia L, Molinari R, Rogers SR, Williams GR, Chiaromonte F, Calovi M. Honey bee colony loss linked to parasites, pesticides and extreme weather across the United States. Sci Rep 2022;12(1):20787.
[26] Dorey JB, Rebola CM, Davies OK, Prendergast KS, Parslow BA, Hogendoorn K, et al. Continental risk assessment for understudied taxa post-catastrophic wildfire indicates severe impacts on the Australian bee fauna. Global Change Biol 2021;27:6551–67.

[27] Latty T, Forster C. Like a moth to a flame: the effect of megafires on pollinators and pollination systems. Curr Opin Insect Sci 2025;68:101304.
[28] Carbone LM, Tavella J, Pausas JG, Aguilar R. A global synthesis of fire effects on pollinators. Global Ecol Biogeogr 2019;28(10):1487–98.
[29] Spigel KM, Robichaud PR. First-year post-fire erosion rates in Bitterroot National Forest, Montana. Hydrol Process 2007;21(8):998–1005.
[30] Robichaud PR, Waldrop TA. A comparison of surface runoff and sediment yields from low- and high-severity site preparation burns. J Am Water Resources Assoc 1994;30(1):27–34.
[31] Benavides-Solorio J, MacDonald LH. Post-fire runoff and erosion from simulated rainfall on small plots, Colorado Front Range. Hydrol Process 2001;15(15):2931–52.
[32] Dove NC, Safford HD, Bohlman GN, Estes BL, Hart SC. High-severity wildfire leads to multi-decadal impacts on soil biogeochemistry in mixed-conifer forests. Ecol Appl 2020;30(4):e02072.
[33] Parks SA, Abatzoglou JT. Warmer and drier fire seasons contribute to increases in area burned at high severity in western US forests from 1985 to 2017. Geophys Res Lett 2020;47(22). e2020GL089858.
[34] Westerling AL, Hidalgo HG, Cayan DR, Swetnam TW. Warming and earlier spring increase western US forest wildfire activity. Science 2006;313(5789):940–3.
[35] Westerling AL. Increasing western US forest wildfire activity: sensitivity to changes in the timing of spring. Philos Trans R Soc Lond B Biol Sci 2016;371(1696):20150178.
[36] Boer MM, Resco de Dios V, Bradstock RA. Unprecedented burn area of Australian mega forest fires. Nature Clim Change 2020;10(3):171–2.
[37] Filkov AI, Cawson J, Swan MH, Penman TD. Wildland fire. In: Meacham BJ, McNamee M, editors. Handbook of Fire and the Environment: The Society of Fire Protection Engineers Series. Cham: Springer; 2022. p. 235–71.
[38] Hanes CC, Wang X, Jain P, Parisien MA, Little JM, Flannigan MD. Fire-regime changes in Canada over the last half century. Can J For Res 2019;49(3):256–69.
[39] McColl-Gausden SC, Bennett LT, Clarke HG, Ababei DA, Penman TD. The fuel–climate-–fire conundrum: how will fire regimes change in temperate eucalypt forests under climate change? Global Change Biol 2022;28(17):5211–26.
[40] Barbero R, Abatzoglou JT, Pimont F, Ruffault J, Curt T. Attributing increases in fire weather to anthropogenic climate change over France. Front Earth Sci 2020;8:104.
[41] Clarke H, Lucas C, Smith P. Changes in Australian fire weather between 1973 and 2010. Int J Climatol 2013;33(4):931–44.
[42] Linley GD, Jolly CJ, Doherty TS, Geary WL, Armenteras D, Belcher CM, et al. What do you mean,'megafire'? Global Ecol Biogeogr 2022;31(10):1906–22.
[43] Lawson DA, Rands SA. The effects of rainfall on plant–pollinator interactions. Arthropod–Plant Interact 2019;13:561–9.
[44] Van Espen M, Williams JH, Alves F, Hung Y, de Graaf DC, Verbeke W. Beekeeping in Europe facing climate change: a mixed methods study on perceived impacts and the need to adapt according to stakeholders and beekeepers. Sci Total Environ 2023;888:164255.
[45] Kornhuber K, Bartusek S, Seager R, Schellnhuber HJ, Ting M. Global emergence of regional heatwave hotspots outpaces climate model simulations. Proc Natl Acad Sci U S A 2024;121(49). p. e2411258121.
[46] Jhawar J, Davidson JD, Weidenmüller A, Wild B, Dormagen DM, Landgraf T, et al. How honeybees respond to heat stress from the individual to colony level. J R Soc Interface 2024;20:20230290.

[47] Chen J, Fisher II A, DeGrandi-Hoffman G, Ozturk C, Smith BH, Fewell JH, et al. Negative effects of excessive heat on colony thermoregulation and population dynamics in honeybees. Ecol Evol Physiol 2026;99(1). Available from https://doi.org/10.1086/739493.

[48] McAfee A, Chapman A, Higo H, Underwood R, Milone J, Foster LJ, et al. Vulnerability of honey bee queens to heat-induced loss of fertility. Nature Sustain 2020;3:367–76.

[49] Stürup M, Baer-Imhoof B, Nash DR, Boomsma JJ, Baer B. When every sperm counts: factors affecting male fertility in the honeybee Apis mellifera. Behav Ecol 2013;24(5):1192–8.

[50] Smoliński S, Langowska A, Glazaczow A. Raised seasonal temperatures reinforce autumn Varroa destructor infestation in honey bee colonies. Sci Rep 2021;11(1):22256.

[51] DeGrandi-Hoffman G, et al. Population growth of Varroa destructor (Acari: Varroidae) in honey bee colonies is affected by the number of foragers with mites. Exp Appl Acarol 2016;69(1):21–34.

[52] Margotta JW, Mancinelli GE, Benito AA, Ammons A, Roberts SP, Elekonich MM. Effects of flight on gene expression and aging in the honey bee brain and flight muscle. Insects 2013;4(1):9–30.

[53] Williams JB, Roberts SP, Elekonich MM. Age and natural metabolically-intensive behavior affect oxidative stress and antioxidant mechanisms. Exp Gerontol 2008;43:538–49.

[54] Rajagopalan K, DeGrandi-Hoffman G, Pruett M, Jones VP, Corby-Harris V, Pireaud J, et al. Warmer autumns and winters could reduce honey bee overwintering survival with potential risks for pollination services. Sci Rep 2024;14(1450):5410.

[55] Rosenzweig C, Iglesius A, Yang XB, Epstein PR, Chivian E. Climate change and extreme weather events: implications for food production, plant diseases, and pests. Global Change Human Health 2001;2(2):90–104.

[56] Calatayud-Vernich P, Calatayud F, Simó E, Picó Y. Pesticide residues in honey bees, pollen and beeswax: assessing beehive exposure. Environ Pollut 2018;241:106–14.

[57] Mullin CA, Frazier M, Frazier JL, Ashcraft S, Simonds R, VanEngelsdorp D, et al. High levels of miticides and agrochemicals in North American apiaries: implications for honey bee health. PLoS ONE 2010;5(3):e9754.

[58] Ostiguy N, Drummond FA, Aronstein K, Eitzer B, Ellis JD, Spivak M, et al. Honey bee exposure to pesticides: a four-year nationwide study. Insects 2019;10(1):13.

[59] Tosi S, Costa C, Vesco U, Quaglia G, Guido G. A 3-year survey of Italian honey bee-collected pollen reveals widespread contamination by agricultural pesticides. Sci Total Environ 2018;615:208–18.

[60] DeGrandi-Hoffman G, Chen Y, Simonds R. The effects of pesticides on queen rearing and virus titers in honey bees (Apis mellifera L.). Insects 2013;4:71–89.

[61] DeGrandi-Hoffman G, Chen Y, Watkins Dejong E, Chambers ML, Hidalgo G. Effects of oral exposure to fungicides on honey bee nutrition and virus levels. J Econ Entomol 2015;108(6):2518–28.

[62] Pettis JS, Lichtenberg EM, Andree M, Stitzinger J, Rose R, van Engelsdorp D. Crop pollination exposes honey bees to pesticides which alters their susceptibility to the gut pathogen Nosema ceranae. PLoS ONE 2013;8(7):e70182.

[63] Goulson D. Review: an overview of the environmental risks posed by neonicotinoid insecticides. J Appl Ecol 2013;50(4):977–87.

[64] Castle SJ, Byrne FJ, Bi JL, Toscano NC. Spatial and temporal distribution of imidacloprid and thiamethoxam in citrus and impact on Homalodisca coagulata populations. Pest Manag Sci 2005;61(1):75–84.

[65] Humann-Guilleminot S, Binkowski ŁJ, Jenni L, Hilke G, Glauser G, Helfenstein F. A nationwide survey of neonicotinoid insecticides in agricultural land with implications for agri-environment schemes. J Appl Ecol 2019;56(7):1502–14.
[66] Tsvetkov N, Samson-Robert O, Sood K, Patel HS, Malena DA, Gajiwala PH, et al. Chronic exposure to neonicotinoids reduces honey bee health near corn crops. Science 2017;356(6345):1395–7.
[67] Williams GR, Troxler A, Retschnig G, Roth K, Yañez O, Shutler D, et al. Neonicotinoid pesticides severely affect honey bee queens. Sci Rep 2015;5(1):14621.
[68] Matzrafi M. Climate change exacerbates pest damage through reduced pesticide efficacy. Pest Manag Sci 2019;75(1):9–13.
[69] Delcour I, Spanoghe P, Uyttendaele M. Literature review: impact of climate change on pesticide use. Food Res Int 2015;68:7–15.
[70] Eigenbrode SD, Adhikari S. Climate change and managing insect pests and beneficials in agricultural systems. Agron J 2023;115(5):2194–215.
[71] Bell JR, Alderson L, Izera D, Kruger T, Parker S, Pickup J, et al. Long-term phenological trends, species accumulation rates, aphid traits and climate: five decades of change in migrating aphids. J Anim Ecol 2015;84(1):21–34.
[72] Hemming D, Bell J, Collier R, Dunbar T, Dunstone N, Everatt M, et al. Likelihood of extreme early flight of Myzus persicae (Hemiptera:Aphididae) across the UK. J Econ Entomol 2022;115(5):1342–9.
[73] Taylor RAJ, Herms DA, Cardina J, Moore RH. Climate change and pest management: unanticipated consequences of trophic dislocation. Agronomy 2018;8(1):7.
[74] Skendžić S, Zovko M, Pajač Živković I, Lešić V, Lemić D. Effect of climate change on introduced and native agricultural invasive insect pests in Europe. Insects 2021;12(11):985.
[75] Savary S, Willocquet L, Pethybridge SJ, Esker P, McRoberts N, Nelson A. The global burden of pathogens and pests on major food crops. Nature Ecol Evolut 2019;3(3):430–9.
[76] Reilly JR, Artz DR, Biddinger D, Bobiwash K, Boyle NK, Brittain C, et al. Crop production in the USA is frequently limited by a lack of pollinators. Proc Biol Sci 1931;287:20200922.
[77] Perrot T, Gaba S, Roncoroni M, Gautier JL, Bretagnolle V. Bees increase oilseed rape yield under real field conditions. Agric Ecosyst Environ 2018;266:39–48.
[78] DeGrandi-Hoffman G, Chambers M. Effects of honey bee (Hymenoptera: Apidae) foraging on seed set in self-fertile sunflowers (Helianthus annuus L.). Environ Entomol 2006;35(4):1103–8.
[79] Rangel J, Tarpy DR. The combined effects of miticides on the mating health of honey bee (Apis mellifera L.) queens. J Apic Res 2015;54(3):275–83.
[80] Tihelka E. Effects of synthetic and organic acaracides on honey bee health: a review. Slov Vet Res 2018;55(2):119–40.

Chapter 10

Systems to generate solutions

Between the fall of 2006 and spring of 2007, beekeepers lost 30%–90% of their colonies from unknown causes. The name given to the undiagnosed affliction leading to the losses was "Colony Collapse Disorder." Since 2007, the beekeeping industry in the United States and other parts of the world have experienced yearly losses that would have been considered extreme in the past, but are now considered routine. In fact, a good year is when colony losses average only 30%. In bad years like 2024–2025, losses can be 60% or more.

Why do we lose so many colonies? Varroa mites certainly are the primary reason for colony losses throughout the world. The act of parasitism is stressful to the bees, but the most serious aspect of Varroa is that while feeding, the mite transmits viruses. Viruses replicate in the bees and, in the case of deformed wing virus, are transmitted to nestmates through mite feeding and also food exchange between nestmates (see refs in [1]). Levels of viruses in adult bees and brood increase as the colony and Varroa populations grow during spring and summer. Other stress factors such as poor nutrition or exposure to pesticides can compromise the immune system of the bees and accelerate increases in virus levels. Varroa populations and virus concentrations peak in the fall just as colonies are preparing to overwinter. Bees that were parasitized during development and are infected with viruses have severely shortened lifespans ([2] also see refs in [1]). If the colony goes into winter with sick and short-lived bees, it will decline and die either during the winter or in the early spring.

Colonies also are lost due to poor nutrition. Locations with flowering plants that can support colonies from spring through fall are increasingly difficult to find. Without good nutrition, colonies cannot rear enough brood to reach population sizes that can survive the winter. Storing sufficient pollen and honey in the hive to survive periods of confinement during the winter might also be difficult if flowering plants are unavailable. The effects of parasites and pathogens also are more pronounced in colonies under nutritional stress because, like any organism, the strength of an immune response depends on good nutrition [3,4]. Colonies that forage on flowering plants have lower virus and Nosema (a pathogen of the honey bee midgut) levels than those fed supplements without pollen [4] and greater expression of biomarkers indicative of colony health [5].

The Honey Bee and Climate Change. DOI: https://doi.org/10.1016/B978-0-443-28995-8.00008-8

Greater access to landscapes with flowering plants is an important part of reducing colony losses, but the plants must be in areas where colonies are not exposed to pesticides. Pesticides can kill colonies outright if exposure is great enough. More often however, colonies are exposed to sublethal pesticide levels. Sublethal exposure to pesticides may not cause colony death outright, but instead will weaken the colony by reducing adult worker lifespan [6], compromising immunity, and altering behavior, motor and cognitive function [7]. Other stress factors such as Varroa or poor nutrition when combined with exposure to pesticides can push the colony past a tipping point and begin the downhill slide to death.

Solutions to nutritional stress and pesticide exposure need to include changes in land use and a sustained commitment to create landscapes that preserve honey bees and other pollinators. Colony Collapse Disorder triggered a rapid growth in public awareness of the importance of pollinators. "Save the Bees" was everywhere. Corporations contributed money for scholarships to train new bee researchers. Governments increased budgets for bee research, and numerous organizations sprang up to support and raise awareness of the importance of pollinators. Seed mixes were developed and pollinator plantings were encouraged. Continued awareness and goodwill by the public are needed to leverage broad and long-term nationwide commitments to improve conditions for honey bees and other pollinators and ensure conditions where populations can thrive.

Even in the absence of stress from Varroa or pesticides, the effects of climate change alone on fall temperatures can cause colony losses during the winter and spring. When temperatures are warm enough for bees to fly late into the fall, and brood rearing has slowed or stopped as it does in temperate regions, the colony can decline just before overwintering. The overwintering bees also will be physiologically older and have reduced lifespans due to the stress associated with flying (see refs in [8]). When colonies resume foraging in the spring, the short-lived overwintered bees can die at rate that are greater than they can be replaced. The colony will decline and could reach a point where it cannot recover. To avoid losses from warm fall temperatures, hives can be moved to indoor cold storage buildings for the winter. Hives also can be taken to southern latitudes where plants bloom and bees can forage and rear brood all winter. Each overwintering strategy has its benefits and challenges, but need to be included in the system for solutions to colony losses.

Solutions to reduce colony losses need to address the major causes, and how climate change is contributing to their severity. In this chapter, I propose systems that could reduce colony losses due to the impacts of climate change on Varroa, bee forage, pesticide exposure, and overwintering conditions. The systems also take into consideration the effects that climate change will have on agriculture and how strategies to control pests need to include pollinator preservation. To implement the solutions will take more than beekeepers adopting the recommendations. Coordination and participation from public and private sectors are needed to improve ecosystems for pollinators and reverse the unsustainable losses of managed honey bee colonies.

10.1 Varroa mites

Climate change has contributed to the harm inflicted by Varroa. Warmer temperatures that cause earlier springs and longer summers prolong the brood rearing period in colonies. Longer periods when colonies rear brood, extend the time when Varroa can reproduce and populations grow. If Varroa populations are high, the brood that is reared in the fall can have multiple Varroa mites enter cells before they are sealed. When this occurs, the emerging adult usually has high virus levels, and a shortened lifespan. Sometimes the bee even exhibits physical deformities caused by the infection such as poorly developed wings typical of deformed wing virus that is transmitted by Varroa [2,9]. Warmer fall temperatures extend periods with flight weather and also the time when Varroa can migrate into colonies on foragers and rapidly increase mite populations [10].

The key to Varroa control is constant vigilance. In the colony, Varroa are constantly on the move between adult bees and sealed brood cells. The actual number of mites in a colony is not easy to determine. During the summer, most of the Varroa population can be under capped cells feeding on pupae. However, sampling adult bees frequently (e.g., every 3 weeks) can provide an estimate of the relative size of the Varroa population and whether it is increasing or remaining stable. Frequent sampling also can catch sharp increases in Varroa numbers. If a sample of adult bees has more than 3–5% with mites, a miticide treatment should be applied [11].

There are numerous miticides to control Varroa that differ in active ingredients and times when they can be used. For example, most miticides should not be applied if colonies are collecting honey that will be extracted and sold. Other miticides should be used only under cooler temperatures or if the colony has little or no uncapped brood. Beekeepers need to vary the miticides they apply to avoid the mites becoming resistant. Another consideration is whether the miticide will accumulate in the wax combs in the hive and create prolonged exposure that can select for resistant mites.

Pesticide analyses of wax comb invariably show that the highest contaminants are miticides (fluvalinate and coumaphos) used for Varroa control [12]. Beeswax is the least renewable resource of the hive because the same combs are used in colonies for years. The persistance of miticides in wax comb can lead to reduced brood survival [13] and negatively affect queen development and emergence [14,15].

Repeated use of the same miticide and its accumulation in wax comb leads to mite resistance [16]. Resistance to synthetic miticides is a large part of why Varroa is so difficult to control. When Varroa was first found in colonies in 1987, a single miticide, fluvalinate, was available for control. In a few years, mites became resistant and colony losses increased [17]. Next, coumaphos was used, but in several years, mites became resistant to that, too [18]. The most recent synthetic miticide for Varroa control is amitraz, but now widespread resistance to that miticide is occurring [19].

Though there are increased incidences of resistance to synthetic miticides, alternatives are available such as organic acids (thymol, formic acid, oxalic acid), and beta plant acids. Except for beta plant acids, organic acids are released as vapors in the colony. Beta plant acids, like the synthetic miticides, kill mites by getting on the body of bees and mites that may be attached there. Applying organic or beta plant acid miticides is a little more labor intensive than inserting a plastic strip coated with a synthetic miticide. The miticides also vary in effectiveness, especially the organic acids and should be applied only within specific temperature ranges (e.g., formic acid: 50–85 °F (10–29.5°C), thymol: 68–77 °F (20–25°C)). Oxalic acid should be used in the fall when there is little or no brood and ambient temperatures range between 50 °F and 60 °F (10°C–15°C). There also are possible side effects especially if used at high ambient temperature. These include interruptions in brood rearing, decreased brood survival, increased adult mortality, decreases in sperm viability in drones and the spermatheca of queens, and queen loss (see refs in [20]). Still, if used correctly, the miticides can keep Varroa at low enough levels to prevent colony loss. The key to keeping the miticides effective against Varroa is to apply different products throughout the year to avoid mite resistance.

In spring and summer, there can be periods when colonies are collecting an abundance of nectar, and storing it as honey. These periods are called "honey flows," and beekeepers often harvest the honey for sale. With the exception of beta plant acids, miticides cannot be applied in colonies during honey flows. Honey production has a long tradition with beekeeping and is a source of revenue. However, honey harvesting might cause colonies to fail later in the season. If mite populations are high in the spring or summer, and miticides are not used because of a honey flow, viruses transmitted by the mites will increase in the colony throughout the summer and fall even if miticides are applied immediately after the honey is harvested and mite populations decline. The viruses will be transmitted among nestmates and brood, causing each brood cycle to have higher virus levels. To prevent colony losses weeks or months after honey flows, colonies should be sampled before the honey flow. Colonies with high mite numbers should be treated and not be available for honey harvest. There might be fewer colonies for honey production but there also could be fewer colonies lost later in the year.

Another way to keep Varroa levels low is to use mite resistant/tolerant bees. There are several genetically selected lines of bees that slow mite reproduction and can keep populations low most of the year [21,22]. However, if queens are lost, a new queen from a mite-resistant/tolerant line needs to be purchased to maintain the resistant traits in the colony. Although mite-resistant lines can keep Varroa levels in check for most the spring and summer, in the fall, mites may migrate into colonies with mite-resistant/tolerant bees just as they do with unselected lines [23]. Only regular sampling at 3–4-week intervals will provide assurances that mite numbers have not exceeded thresholds where treatments are required.

10.2 Increased availability of flowering plants

The second component in the system to produce healthy robust colonies and reduce losses focuses on nutrition. Habitat loss has been a major contributor to declines in pollinator populations (see refs in [24]) and to nutritional stress that leads to reduced growth and survival of managed colonies. Landscapes with numerous species of flowering plants blooming from spring until fall need to be made available to beekeepers. Continuous flowering ensures that pollen and nectar are available for colonies while they grow in spring and summer and prepare for winter in the fall. Although beekeepers can feed colonies protein supplement when pollen is not available, the diets are not as effective for rearing brood and preventing disease as having bees foraging on flowering plants [4,25].

Providing flowering plants especially for managed colonies should begin in early spring with areas around almond orchards. Each year almost 2 million honey bee colonies are moved to California to pollinate the more than 1 million bearing acres of almonds [26]. Almonds bloom in February and coincides with the start of egg laying and brood rearing in winter. Before almond bloom, flowering plants can be sparse, and colonies are kept alive and rearing brood before bloom using protein or pollen supplements.

A solution to support colonies that are pollinating almonds in California is planting early season pollinator forage such as mixtures of spring mustards (e.g., *Brassica ruvo*, *Sinapis alba*, *Brassica juncea Nemfix*, *Raphanus sativus*) [27] (Fig. 10.1). Mustards and other plants that bloom early in the spring can provide colonies with pollen sources before, during, and after almond bloom in California. Plants that flower before almond bloom provide a much-needed source of pollen to colonies brought to almond orchards before trees come into bloom. The colonies are rearing brood and building populations and incoming pollen is needed. The mustards also bloom while almond trees are flowering and supply an alternative pollen source that can provide the diverse pollen sources needed to fulfill nutritional requirements of colonies. Mustards continue to flower after almond bloom and provide colonies with pollen so that their populations can continue to build.

A study was conducted to measure the effects of planting mustard between rows of trees in almond orchards on colony size after almond bloom. Colonies were brought into orchards with and without the mustard plantings. The size of the colonies was estimated in January just prior to almond bloom until October when the colonies were preparing for overwintering. The study

FIGURE 10.1 Flowering spring mustards (Brassica spp.) in California, USA, almond orchards. The flowering plants provide nectar and pollen before, during, and after almond bloom to honey bee colonies brought to the orchards for almond pollination.
Photograph from California State University, Chico, Center for Regenerative Agriculture and Resilient Systems.

revealed long-term benefits of planting mustard mixes in almond orchards. Colonies that could forage on mustard flowers beginning in January were larger for the remainder of the season compared with those from orchards without the mustard plantings [27]. The study demonstrates the importance of early spring flowering sources to maximize colony build up early in the season and create momentum in positive feedback loops that can carry on throughout the remainder of the year.

Planting pollinator forage can have benefits for almond growers that extend beyond healthier colonies for pollination. Mustards and other pollinator planting in orchards grow during the winter rainy season when almond trees are dormant. The roots of the pollinator plantings can increase water infiltration into the soil and reduce runoff and erosion. The plantings also can enhance the colonization of microbes especially fungi (arbuscular mycorrhizae) and can improve the ability of almond tree roots to absorb nutrients and water [28]. Although many benefits of planting pollinator forage in almond orchards have been found, there is still much to learn about the effects of plants in the understory of orchards as a refuge for pest insects. Whether the plantings interfere with frost management also needs more study.

Whether colonies are used for almond pollination or are becoming active again after overwintering in apiaries, spring, and summer forage is essential for building colony populations. In the United States, there is great potential for expanding landscapes to support all species of pollinators. In 2015, the U.S. Pollinator

Research Action Plan (https://obamawhitehouse.archives.gov/sites/default/files/microsites/ostp/Pollinator%20Research%20Action%20Plan%202015.pdf) recommended restoring or enhancing 7 million acres of land for pollinators over the next 5 years through Federal actions and public/private partnerships. Although the goal has not yet been met, it is achievable. Only about 5% of the land mass in the United States is covered by cities, towns, and roads. About 40% of the country is public land (i.e., National and State Parks and Forests, wildlife refuges, etc.) and managed by federal, state, and local governments (914 million acres or 3.7 million km^2). There also are large areas of crop land that can be made available to beekeepers through the Conservation Reserve Program (CRP). CRP is a federal program where landowners are paid an annual rent to remove environmentally sensitive land from crop production and instead plant vegetative covers to prevent erosion, improve water quality, and conserve wildlife habitats (https://www.nrcs.usda.gov/programs-initiatives/conservation-reserve-program). The CRP land can be planted to seed mixtures that flower and attract honey bees and other pollinators. The fields also can be nesting sites for wild (i.e., unmanaged) bees. Benefits have been documented in numerous studies showing seeding CRP and public land with pollinator-attractive native plants can produce abundant and diverse floral landscapes that attract and support numerous species of foraging bees [5,29–31].

Expansion of CRP land especially in the Midwest and Northern Great Plains of the United States (i.e., regions in Montana, Wyoming, North and South Dakota, Minnesota) is particularly important to U.S. beekeepers since millions of colonies summer there and suitable locations for apiaries are hard to find. The region had historically provided vast expanses of open land with flowering plants where honey bee colonies could thrive and beekeepers were assured yearly honey crops. However, there has been a steady reduction in acreage containing bee forage and an increase in plantings of biofuel crops (corn and soybean) (see refs in [32]). At first glance, it seemed as if biofuels would lessen fossil fuel use and be part of the solution for reducing carbon emissions. This is not the case. Corn and soybeans require large tracks of land to produce enough material to convert into biofuel. More than 30 million acres of corn and 10 million acres of soybeans are used in the production of ethanol, but even with this large land investment, corn supplies about 4% and soybeans less than 1% of transportation fuels used in the United States (World Resource Institute, https://www.wri.org/insights/increased-biofuel-production-impacts-climate-change-farmers#main-content). Furthermore, producing and refining biofuels create their own carbon emissions. The expansion of corn and soybean plantings has created vast monocultures with crops that do not provide nutritious forage for bees and contributes to the decline in pollinator populations.

A solution that could increase flowering plants for honey bees and other pollinators is planting cover crops. Typically, cover crops are rye grasses and oats but can also be winter oilseed crops such as pennycress (*Thlaspi arvense* L.), camelina (*Camelina sativa* L.), and canola (*Brassica napus* L.) (Fig. 10.2).

FIGURE 10.2 Canola (*Brassica napus* L.) (A), pennycress (*Thlaspi arvense* L.) (B), and camelina (*Camelina sativa* L.) (C) used as cover crops for pollinators. Examples of cover crops that can be planted as pollen and nectar sources for pollinators.

Cover crops are planted during the off season of cash crops or in fallow fields to prevent erosion, improve water quality and soil health through nutrient recycling, and suppress the spread of weeds (see refs in [33]). In parts of the Northern Great Plains called the Northern Corn Belt (southern Wisconsin, Minnesota, and South Dakota), the growing season for corn begins in May and ends in October. The short growing season causes fields to be vulnerable to erosion for over half the year. Planting winter oilseed crops can provide green cover to open fields. The crops bloom from late April to early June and can provide early season forage for honey bee colonies after pollinating orchards in the western United States as well as for wild bees emerging from hibernation. Winter oilseed crops such as pennycress, camelina, and canola also can generate revenue because the seed contains valuable oils that can be sold and incorporated in a variety of products such as biofuels, feedstock, cosmetics, and cooking oils [32].

Preserving uncultivated land for planting cover crops and pollinator seed mixtures is particularly important for relieving nutritional stress many colonies experience during and after pollinating crops. Although orchards and fields in bloom may seem like a bonanza of nectar and pollen, floral diversity is imperative to assure that nutritional needs of colonies are being met (see refs in [27]). Areas in field margins, fence lines, and uncultivated patches of land can be planted with cover crops or pollinator seed mixtures to enhance diversity of pollen sources and also serve as refuges and nesting sites for wild bees [34]. Projects to enhance floral resources and reverse pollinator losses are underway in many countries (e.g., United Kingdom, Sweden, Switzerland, and the United States). Researchers are developing cover crops, hedgerows, and wildflower strips around field margins and interspersed with crop rows to support pollinators and other beneficial insects (see refs in [33]). Exchanging information among the projects will accelerate their implementation and provide much-needed information on cover crops wildflower mixtures that provide the most benefits to pollinators and growers.

A caveat to planting cover crops for bees in fields where corn, soybeans, cotton, wheat, and other crops were planted is the presence of neonicotinoids in the soil. Neonicotinoids are neurotoxins that can effectively control a wide range of pest insects that either consume plant parts or are sap feeding piercing–sucking insects (e.g., aphids (Family Aphididae)), whiteflies (Family Aleyrodidae), or planthoppers (e.g., Families Delphacidae, Flatidae, Fulgoridae). Neonicotinoids are systemic insecticides, meaning that they are absorbed by the roots and leaves of plants and distributed among plant tissues. Plants take up less than 1%–5% on the neonicotinoid mass applied within seed coatings. The remainder of the pesticide can accumulate in the soil, or move to field margins (see refs in [35]). The half-lives of neonicotinoids can range from 191 to 545 days for clothianidin in soil depending on the formulation and soil type (see refs in [36]). Neonicotinoids have high water solubility, which facilitates their uptake by roots and transport to plant tissues above the ground, including flowers, nectar, and pollen. The water solubiliy also causes the compounds to move laterally through the soil. Consequently, neonicotinoids have been detected in non-crop plants on the margins of treated fields. The compounds also were detected in cover crops grown in fields previously planted with neonicotinoid-treated seed. A study in Belgium detected neonicotinoids in soils and floral resources of cover crops planted in fields where neonicotinoid seeds had not been planted for the past 3 years [36]. Neonicotinoids are harmful to honey bee colonies, causing reduced adult bee longevity and queen loss (see [37,38]). The concentrations of neonicotinoids depended on the type of cover crop, therefore pollen and nectar from cover crops planted in fields previously planted with neonicotinoid-treated seed should be analyzed for the presence and concentration of this pesticide before exposing colonies to these sites. The harm that could occur to wild bees also is a concern and reason to find alternatives for this class of pesticides.

10.3 Reducing pesticide use

Solutions to reduce colony losses must include more judicious use of pesticides. Honey bees are exposed to pesticides while foraging agricultural crops, but there are many other ways that honey bees and other pollinators can encounter these toxins. As described above, neonicotinoids that were applied as seed treatments can move laterally through the soil to the margins of field where they can be detected in the plants growing there. Honey bees also are exposed to neonicotinoids while treated seeds are being planted. For instance, the abrasion of corn seeds during planting generates dust that is laden with the insecticide. The dust disperses into the environment and can contaminate water, soil, and vegetation. Residues of seed treatment dust containing the neonicotinoids, clothianidin, and thiamethoxam, were detected in pollen collected by bees in apiaries surrounded by fields where neonicotoid treated corn seed was planted [39]. The problem is pervasive across the United States as 79%–100% of corn is grown from seed treated with neonicotinoids [40]. In many instances, the treated seed is used as an "insurance policy" for losses from insect pests that might occur rather than to target a specific pest.

Honey bees often are exposed indirectly to insecticides, but can come into direct contact with others such as fungicides. There are many types of fungicides with different modes of action, but they are formulated to be long-acting either through slow-release granules or by being absorbed by the plant as a systemic compound. Because fungicides have no immediate toxicity to bees at levels encountered in the field, they can be applied while colonies are in blooming orchards and fields to pollinate crops. However, studies show that fungicide exposure is correlated with poor brood rearing, reduced pollen consumption and digestion, increased presence of disease, an inability to rear queens to emergence, and colony loss (see refs in [41]). Because of direct exposure of bees to fungicides, it is not surprising that these compounds are some of the most commonly found in pollen and wax comb [12,42].

An analysis of trapped pollen from incoming foragers, bee bread stored in colonies, and wax comb revealed multiple pesticides in more than 90% of the samples. The wax comb samples averaged 8 and as many as 31 different pesticides [12]. The comb is an archive of pesticide use in the United States where compounds like dichloro-diphenyl-trichloroethane that have long been banned in the United States are still detected in wax comb. The compounds are in the comb not necessarily because they have been recently applied, but because they persist in the environment.

Reducing pesticide use is a difficult and complicated solution for reducing colony losses. Pesticides play an important role in producing the crops needed to feed a growing world population. Consumers also have expectations on the quality and price of food. However, current agricultural practices are striking an uneven balance between agricultural production and care for the environment. Biodiversity in general and pollinators in particular are being impacted and soil, water, and air are suffering long-term contamination. With warming temperatures expected to increase agricultural pests and present new challenges to crop production, environmental imbalances could increase. Furthermore, pesticides are largely petroleum products whose production directly contributes to climate change.

Current agricultural practices need to be modified to prevent further environmental harm and reduce stress from pesticides on honey bee colonies. New ways to prevent crop losses from diseases and pests also need to be developed. These include biological plant biostimulants (e.g., microbes, humic acids, or seaweed extracts) that stimulate natural processes of plants and enhance nutrient uptake, plant immune responses, tolerance to insect feeding, plant growth, and crop quality and yield. Another way to reduce pesticide use is through widespread implementation of integrated pest management (IPM). IPM combines biological, physical, cultural, and chemical tools to keep pest population below levels where they cause economic damage to the crop. Successful IPM programs also can change the perception that applying pesticides is the only way to prevent crop losses. Agroecological and organic practices that use ecological principles to design and maintain sustainable food systems without the use of synthetic pesticides also need continued development. Finally, technology-driven approaches such as precision agriculture could reduce pesticide use especially on large acreages by identifying specific locations in fields where inputs such as water, fertilizer, and pesticides are needed [43].

Although reducing the use of pesticides will require growers to adopt new management strategies, there are benefits that could provide motivation. Fewer applications of pesticides and fertilizers lower the costs of crop production and the risks that exposure to these chemicals have on human health. The occurrence of pest resistance also is reduced when pesticides are applied less often. Improved soil health, less runoff of pesticides into ground water, and larger populations of beneficial insects including predators, parasites, and pollinators also are benefits of reduced pesticide use. These benefits could reshape agroecosystems and make them more sustainable as climactic conditions become more unstable and affect the livelihood of growers and beekeepers alike.

10.4 Overwintering

Of all the management adjustments that need to be made to colony losses from climate change, perhaps overwintering needs the most attention. Due to climate change, winter conditions can cause colony loss even if Varroa is controlled and bees are well fed. Increasing incidence of extended summers and warm falls have revealed the critical role temperature may play in colony preparation for overwintering particularly in temperate climates. Warmer fall temperatures also have exposed how little we know about the transition from summer to winter bees, and the role of temperature in that transition.

Colonies that spend the summer in temperate areas begin preparing for winter in late summer. Worker bees start storing nutrients in their fat bodies and reduce the amount of brood they rear. The process continues throughout the fall in preparation for confinement in the hive during the cold of winter. Research to determine how colony populations prepare for winter and transition from summer to winter bees was done in northern latitudes where low fall temperatures and deceasing daylength were followed by anatomical and physiological changes in worker bees [44]. The changes were interpreted as transitions from summer to winter bees. Conditions that were not previously considered, perhaps because they did not commonly occur, was whether the transition from summer to winter bees is affected when higher fall temperatures prolong foraging and brood rearing. Do warmer fall temperatures cause honey bees to be ill-prepared for periods of confinement in the hive during the winter? This knowledge gap is important because knowing the conditions that affect the transition from summer to winter bees and whether the transition has occurred in a given year will determine how to manage colonies to avoid overwintering losses.

Apiaries in lower elevations in temperate latitudes might have years with cold fall temperatures, but predictions are that they will occur less frequently. By mid-century, some areas that now are suitable for successful overwintering will have years where warm periods in fall and winter will make it difficult for colonies to survive [8]. Colonies that spend the summer in these areas may need to be moved to more northern latitudes or higher elevations in the fall to ensure successful overwintering.

An alternative to overwintering colonies in apiaries is to move hives into indoor cold storage buildings for overwintering. Cold storage will stop both late fall foraging and Varroa migration. However, to successfully overwinter in cold storage, colonies must contain winter bees. Colonies that spend the summer in southern latitudes and are rearing brood late in the fall do not have bees that can survive long periods of confinement in cold storage [45]. The time when colonies that summer in temperate regions have completely transitioned from summer to winter bees might differ among years depending on conditions in the fall. Colonies may transition to winter bees later during warm falls and earlier when fall temperature are low.

Some colonies that spend the summer in temperate areas like the Northern Great Plains of the United States are taken to California for overwintering. This is a common practice for beekeepers that rent their colonies for almond pollination. In years with winter rains in California, there may be enough forage that colonies avoid nutritional stress and perhaps even rear some brood throughout the winter. However, in years with dry winters, bees will forage but flowering plants may be sparse. Colonies can dwindle throughout the winter if they do not rear enough brood to replace the bees that die while foraging. To maintain colony populations, bees have to be fed pollen patties or protein supplements throughout the winter so that brood rearing can occur. Varroa levels also need to be monitored at frequent intervals if colonies are active throughout the winter. While the bees are flying, Varroa can be migrating on foragers and spread throughout apiaries. To avoid losing colonies to Varroa, bees will need to be sampled at 3–4-week intervals and treated as necessary.

Another overwintering option is to move colonies from their summer apiaries to southern latitudes. In some areas, flowering plants such as *Solidago* spp., *Aster* spp., and *Baccharis halmifolia* can supply pollen in October and November. In winter, *Alnus serrulata*, *Acer rubrum*, *Lamium* spp., and *Ulmus* spp. can provide bees with flowers to forage and pollen to rear brood [46]. However, since bees are foraging and rearing brood, colonies need to be sampled for Varroa throughout the winter and treatments applied as necessary. If colonies are overwintered in states along the Gulf Coast, such as Mississippi, Alabama, Louisiana, Texas, and Florida, apiary locations need to be chosen carefully. The National Oceanic and Atmospheric Administration predict an average sea level rise of around 10 to 12 in. (0.25–0.3 m) by 2050 across the U.S. coastlines (Sea Level Rise and Coastal Flooding Impacts, https://coast.noaa.gov/slr/). Areas along the Gulf of Mexico are projected to experience significant effects from sea level rise and increased incidence of high-tide flooding.

10.5 Final thoughts

As in many areas of agriculture, climate change is affecting the beekeeping industry and requiring that methods for colony management evolve to fit the times. Temperature ranges, rainfall amounts, and the frequency of severe weather events are less predictable in terms of when and where they will occur, and their effects can be widespread and complex. Maintaining healthy honey bee colonies is possible but may require more labor and materials than beekeepers currently are expending. Costs of keeping colonies alive will increase and the added expenses will generate higher prices for colony rental, honey, and crops requiring honey bee pollination.

Although many of the solutions I propose can be adopted by beekeepers, others such as reducing pesticide applications, planting cover crops and creating landscapes that support pollinator populations will require broad cooperation across the agricultural community. Expanding habitats to support pollinators and opening them to beekeepers will require conversations among public and private land managers, policymakers, and beekeepers. Decisions on where apiaries can be established need to be based on scientific evidence particularly concerning interactions between wild bees and managed honey bee colonies. The effects of climate change on honey bees and the beekeeping industry are just being realized, but because of the irreplaceable contributions of honey bees to ecosystem stability, food production, and human health, there needs to be a sense of urgency in addressing them.

References

[1] Grozinger CM, Flenniken ML. Bee viruses: ecology, pathogenicity, and impacts. Annu Rev Entomol 2019;64:205–26.

[2] Dainat B, Evans JD, Chen YP, Gauthier L, Neumann P. Dead or alive: deformed wing virus and Varroa destructor reduce the life span of winter honeybees. Appl Environ Microbiol 2012;78(4):981–7.

[3] Alaux C, Ducloz F, Crauser D, Le Conte Y. Diet effects on honeybee immunocompetence. Biol Lett 2010;6(4):562–5.

[4] DeGrandi-Hoffman G, Chen Y, Rivera R, Carroll M, Chambers M, Hidalgo G, et al. Honey bee colonies provided with natural forage have lower pathogen loads and higher overwinter survival than those fed protein supplements. Apidologie 2016;47(2):186–96.

[5] Ricigliano VA, Mott BM, Maes PW, Floyd AS, Fitz W, Copeland DC, et al. Honey bee colony performance and health are enhanced by apiary proximity to US Conservation Reserve Program (CRP) lands. Sci Rep 2019;9(1):4894.

[6] Wu JY, Anelli CM, Sheppard WS. Sub-lethal effects of pesticide residues in brood comb on worker honey bee (Apis mellifera) development and longevity. PLoS ONE 2011;6(2):e14720.

[7] Chmiel JA, Daisley BA, Pitek AP, Thompson GJ, Reid G. Understanding the effects of sublethal pesticide exposure on honey bees: a role for probiotics as mediators of environmental stress. Front Ecol Evolut 2020;8:22.

[8] Rajagopalan K, DeGrandi-Hoffman G, Pruett M, Jones VP, Corby-Harris V, Pireaud J, et al. Warmer autumns and winters could reduce honey bee overwintering survival with potential risks for pollination services. Sci Rep 2024;14(1450):5410.

[9] Floris I, Pusceddu M, Satta A. How the infestation level of Varroa destructor affects the distribution pattern of multi-infested cells in worker brood of Apis mellifera. Vet Sci 2020;7(3):136.

[10] DeGrandi-Hoffman G, Ahumada F, Zazueta V, Chambers M, Hidalgo G, Watkins deJong E. Population growth of Varroa destructor (Acari: Varroidae) in honey bee colonies is affected by the number of foragers with mites. Exp Appl Acarol 2016;69(1):21–34.

[11] Gregorc A, Sampson B. Diagnosis of varroa mite (Varroa destructor) and sustainable control in honey bee (Apis mellifera) colonies—a review. Diversity 2019;11(12):243.

[12] Mullin CA, Frazier M, Frazier JL, Ashcraft S, Simonds R, vanEngelsdorp D, et al. High levels of miticides and agrochemicals in North American apiaries: implications for honey bee health. PLoS ONE 2010;5(3):e9754.

[13] Medici SK, Castro A, Sarlo EG, Marioli JM, Eguaras MJ. The concentration effect of selected acaricides present in beeswax foundation on the survival of Apis mellifera colonies. J Apic Res 2012;51(2):164–8.

[14] Haarmann T, Spivak M, Weaver D, Weaver B, Glenn T. Effects of fluvalinate and coumaphos on queen honey bees (Hymenoptera: Apidae) in two commercial queen rearing operations. J Econ Entomol 2002;95(1):28–35.

[15] Collins AM, Pettis JS, Wilbanks R, Feldlaufer MF. Performance of honey bee (Apis mellifera) queens reared in beeswax cells impregnated with coumaphos. J Apic Res 2004;43(3):128–34.

[16] Medici SK, Maggi MD, Sarlo EG, Ruffinengo S, Marioli JM, Eguaras MJ. The presence of synthetic acaricides in beeswax and its influence on the development of resistance in Varroa destructor. J Apic Res 2015;54(3):267–74.
[17] Milani N. The resistance of Varroa jacobsoni Oud. to acaricides. Apidologie 1999;30(2-3):229–34.
[18] Mitton GA, Meroi Arcerito F, Cooley H, Fernandez de Landa G, Eguaras MJ, Ruffinengo SR, et al. More than sixty years living with Varroa destructor: a review of acaricide resistance. Int J Pest Manag 2022:1–18.
[19] Rinkevich FD. Detection of amitraz resistance and reduced treatment efficacy in the Varroa Mite, Varroa destructor, within commercial beekeeping operations. PLoS ONE 2020;15(1):e0227264.
[20] Tihelka E. Effects of synthetic and organic acaracides on honey bee health: a review. Slov Vet Res 2018;55(2):119–40.
[21] Mondet F, Beaurepaire A, McAfee A, Locke B, Alaux C, Blanchard S, et al. Honey bee survival mechanisms against the parasite Varroa destructor: a systematic review of phenotypic and genomic research efforts. Int J Parasitol 2020;50(6–7):433–47.
[22] Scaramella N, Burke A, Oddie M, Dahle B, de Miranda JR, Mondet F, et al. Host brood traits, independent of adult behaviours, reduce Varroa destructor mite reproduction in resistant honeybee populations. Int J Parasitol 2023;53:565–71.
[23] DeGrandi-Hoffman G, Ahumada F, Danka R, Chambers M, Watkins deJong E, Hidalgo G. Population growth of Varroa destructor (Acari: Varroidae) in colonies of Russian and unselected honey bee (Hymenoptera: Apidae) stocks as related to numbers of foragers with mites. J Econ Entomol 2017;110(3):809–15.
[24] Goulson D, Nicholls E, Botías C, Rotheray EL. Bee declines driven by combined stress from parasites, pesticides, and lack of flowers. Science 2015;347(6229):1255957.
[25] Tsuruda JM, Chakrabarti P, Sagili RR. Honey bee nutrition. Vet Clin N Am Food Anim Pract 2021;37(3):505–19.
[26] CDFA. In: USDA, NASS (Eds.). California Almond Acreage Report; 2022.
[27] Mayack C, Carlson M, Niño BD, Niño EL, Seshadri A. Impacts of almond pollination service and inter-row cover cropping on honey bee colony strength and performance. Sci Total Environ 2025;983:179703.
[28] Wauters VM, Jarvis-Shean K, Williams N, Hodson A, Hanson BD, Haring S, et al. Developing cover crop systems for California almonds: current knowledge and uncertainties. J Soil Water Conserv 2023;78(1):5A–11A.
[29] Quinlan GM, Milbrath MO, Otto CR, Isaacs R. Farmland in US Conservation Reserve Program has unique floral composition that promotes bee summer foraging. Basic Appl Ecol 2021;56:358–68.
[30] Dolezal AG, St. Clair AL, Zhang G, Toth AL, O'Neal ME. Native habitat mitigates feast–famine conditions faced by honey bees in an agricultural landscape. Proc Natl Acad Sci U S A 2019;116(50):25147–55.
[31] Williams NM, Ward KL, Pope N, Isaacs R, Wilson J, May EA, et al. Native wildflower plantings support wild bee abundance and diversity in agricultural landscapes across the United States. Ecol Appl 2015;25(8):2119–31.
[32] Eberle CA, Thom MD, Nemec KT, Forcella F, Lundgren JG, Gesch RW, et al. Using pennycress, camelina, and canola cash cover crops to provision pollinators. Ind Crop Products 2015;75:20–5.
[33] Bryan CJ, Sipes SD, Arduser M, Kassim L, Gibson DJ, Scott DA, et al. Efficacy of cover crops for pollinator habitat provision and weed suppression. Environ Entomol 2021;50(1):208–21.

[34] Nicholls CI, Altieri MA. Plant biodiversity enhances bees and other insect pollinators in agroecosystems. Agron Sustain Dev 2013;33(2):257–74.

[35] Morrison BA, Xia K, Stewart RD. Evaluating neonicotinoid insecticide uptake by plants used as buffers and cover crops. Chemosphere 2023;322:138154.

[36] Tarano IG, Boumal T, De Toffoli M, Buron M, Kiljanek T, Martel AC, et al. Are cover crops a potential threat for pollinators due to clothianidin residues in floral resources? Environ Monit Assess 2025;197(11):1260.

[37] Tsvetkov N, Samson-Robert O, Sood K, Patel HS, Malena DA, Gajiwala PH, et al. Chronic exposure to neonicotinoids reduces honey bee health near corn crops. Science 2017;356(6345):1395–7.

[38] Williams GR, Troxler A, Retschnig G, Roth K, Yañez O, Shutler D, et al. Neonicotinoid pesticides severely affect honey bee queens. Sci Rep 2015;5(1):14621.

[39] Lin CH, Sponsler DB, Richardson RT, Watters HD, Glinski DA, Henderson WM, et al. Honey bees and neonicotinoid-treated corn seed: contamination, exposure, and effects. Environ Toxicol Chem 2021;40(4):1212–21.

[40] Douglas MR, Tooker JF. Large-scale deployment of seed treatments has driven rapid increase in use of neonicotinoid insecticides and preemptive pest management in US field crops. Environ Sci Technol 2015;49(8):5088–97.

[41] Fisher II A, DeGrandi-Hoffman G, Liao LH, Tadei R, Harrison JF. The challenge of balancing fungicide use and pollinator health. Adv Insect Physiol 2023;64:117–90.

[42] Rondeau S, Raine NE. Fungicides and bees: a review of exposure and risk. Environ Int 2022;165:107311.

[43] He Q, Sun Y, Yi M. Evolutionary game of pesticide reduction. Management for Sustainable Agriculture: an analysis based on local governments, farmers, and consumers. Sustainability 2023;15(12):9173.

[44] Mattila HR, Harris JL, Otis GW. Timing of production of winter bees in honey bee (Apis mellifera) colonies. Insectes Sociaux 2001;48(2):88–93.

[45] DeGrandi-Hoffman G, Corby-Harris V, Graham H, Watkins deJong E, Chambers M, Snyder L. The survival and growth of honey bee (Hymenoptera: Apidae) colonies overwintered in cold storage: the effects of time and colony location. J Econ Entomol 2023;116(4):1078–90.

[46] DeGrandi-Hoffman G, Graham H, Corby-Harris V, Chambers M, Watkins-deJong E, Ihle K, Bilodeau L. Adapting overwintering honey bee (Apis mellifera L.) colony management in response to warmer fall temperatures associated with climate change. Insects 2025;16(3):266.

Chapter 11

Final thoughts: individual versus community in addressing climate change

We are here to awaken from the illusion of our separateness

–Thich Nhat Hanh

We began this book from the broad perspective of systems thinking and how outcomes are products of connections among system components. The chapters that followed discussed honey bee colonies and how climate change affects them using a systems framework. My final thoughts take a wide view of climate change and the systems that produced it. The systems include human components derived from our mind sets, perceptions of separateness, and even our evoltuonary history. Earlier I shared that from a systems perspective, there are no isolated incidents, just outcomes from a series of interconnected events. Complex systems can generate a range of outcomes with some having a greater chance of occurring than others. Some outcomes occur over short periods of time while others require longer periods of time to play out.

Climate change is an excellent example of an outcome that needed time to occur. The system was put in motion in the mid-19th century when burning coal to fuel the Industrial Revolution caused increasing amounts of greenhouse gases to be released into the atmosphere. Over time, greenhouse gas emissions accumulated in the atmosphere and reached levels that exceeded the Earth's systems to sequester them. Cities where coal was burned for manufacturing and to heat homes were the first to experience the air pollution. In 19th century Britain, where the growth of coal burning fueled the Second Industrial Revolution, the air in London combined with the dense fog was so polluted with particulates that respiratory illnesses especially bronchitis became prevalent among the population. So severe was the pollution that mortality rates associated with bronchitis rose in London from 25 deaths per 100,000 inhabitants per year in 1840 to nearly 300 deaths in 1890. Essentially, 1 person in 350 living in London in 1890 died from bronchitis [1].

The system of burning fossil fuels like coal has expanded globally and has been running for more than 150 years. The product of this long-running system

The Honey Bee and Climate Change. DOI: https://doi.org/10.1016/B978-0-443-28995-8.00012-X

is greenhouse gas levels that are causing rising air and ocean temperatures, and jeopardizing human health and ecosystem survival on a global scale. The warmest years on record across the globe have occurred since 2001, with 2024 being the warmest year since recording began in 1850 [2].

So how did we get to where we are today in terms of climate change? Walking back to the origins of climate change, we need to begin by understanding the relationship between humans and nature. I would submit that the beginnings lie in the ideology that humans are separate from nature and have a greater intrinsic value than other species. This ideology is called anthropocentrism. Where did that ideology originate? Some philosophers and ethicists consider the separation of humans from nature to be a deep human tendency that pre-dates written history. Anthropocentrism can be found in ancient creation stories (e.g., Judeo-Christian Book of Genesis; 1450–1400 BCE) and the writings of ancient philosophers. For example, Aristotle (384–322 BCE) defined humans as rational animals and all other life forms as sensory (*Politics* and *Nicomachean Ethics*). The separation of humans from the natural world became further entrenched in human consciousness by the philosophy of Rene Descartes (1596–1650). As a mathematician and philosopher, Descartes proposed that nature was based on mathematical principles, and that nature could be defined in mechanical terms. Understanding the mathematical principles on which nature was built would allow humans to master the natural world. Taking it a step further, Descartes proposed that nature, living or inert, is composed of objects to fulfill humanity's needs. Animals were considered to be machines that could be manipulated by understanding their mechanics. Nature was viewed as an infinite resource that humans could use, often to fuel economic growth, without any qualms or long-term repercussions.

Internalizing Descartes' view of the natural world created a utilitarian relationship with the natural world and led to the development of systems for gathering natural resources with increasing efficiency. Systems were constructed to mine and burn fossil fuels that, in turn, supported transportation and to create energy systems for generating heat and later electricity. The belief that nature is an endless resource from which we are disconnected provided an ideal philosophical framework for the high-pitched fervor for economic growth and improved standards of living. Put into practice that is still occurring today, the utilitarian relationship with nature provided the rationale for destroying mountains to mine coal and minerals, cutting down 95% of the original 2 million acres of Pacific redwood (*Sequoia sempervirens*) forests during the 1850s to the early 20th century, plowing millions of acres of native grasslands and oblitering Bison populations in the Great Plains of the United States (1909–29) and since 1978, destroying up to 20% of the Amazon Rain Forest to expand agricultural lands for crops and cattle.

The disconnection between humans and nature caused species extinctions and environmental disasters in the past and today might be contributing to the denial by some individuals that humans are complicit in altering the earth's climate. In the United States, about a quarter of the people surveyed by the Pew Research Trust believe that global warming in caused primarily by natural weather patterns and that humans have little role. Another 14% do not believe that there is evidence that the planet is warming, even though the impacts of human activity on global warming have been documented on every continent, in every ocean, and in a majority of plant and animal species [3].

Assuming that humans were separate from nature and, perhaps the unspoken understanding that organisms were separate from each other, persisted until it was challenged at least in the Western world, by the writings of the British botanist, Sir Arthur George Tansley. In 1935, Tansley introduced the concept of the ecosystem [4]. This concept describes the interactions that individual species have among themselves, with other organisms living in the same environment, and with the physical environment (e.g., soil, air, climate). Early studies focused on single interactions between species such as predator–prey or parasitism [5]. Further observations found that various species are commonly found together because they have similar habitat requirements. Taking a broader view of relationships among species found them to be more nuanced than predator prey. For example, relationships can be mutually beneficial such as those between pollinators and flowering plants. Furthermore, the relationships can extend between and among different species at multiple levels (e.g., macro- and microscopic) so that ecosystems are more accurately considered to be communities. The biology of any one species needs to be considered in the context of its relationships with other members of the community where it lives. The interactions among species influence all aspects of the community either directly or indirectly through downstream effects on community members.

Communities can be seen in a systems framework where subsystems of individual species are interconnected with subsystems of other species. The greater the number of species in an ecosystem (i.e. biodiversity), the more connections can be established among the subsystems. Communities with high biodiversity have multiple connections among its members, and are better equipped to regain stability after experiencing disruptions than those with less species diversity. The multiple interactions between and among species also can create a positive feedback loop that helps maintain biodiversity [5].

How biodiverse communities can be sustained following a disturbance might be made clearer using an example of a community with multiple species of bees and flowering plants. The foraging subsystem in a honey bee colony produces foraging bees that visit flowers for nectar and pollen. The foraging subsystem is connected to the floral subsystem in plants that produces visual and odor cues to attract bees and nectar and pollen for the bees to collect. The product of the interactions between honey bee and plant subsystems is pollination and seed set. Biodiverse communities can have multiple species of flowering plants that can support numerous bee species that forage flowers and cross-pollinate them. If environmental conditions such as those that cause poor overwintering survival limit honey bee populations in a given season, the connections between other species of bees and the flowering plants are still available. The bees will forage and pollinate the flowers so that seed set can occur. In the process, the diversity of flowering plants and the bee species in the community are both preserved.

The interactions among species in an ecosystem can be seen from a behavioral perspective, but there are other ways that connections among organisms may occur. Shortly after Tansley's publication on ecosystems, Raymond Lindeman (1942) introduced the concept that ecosystems have energy flows and nutrient cycling among species. From studies in aquatic ecosystems, Lindemann introduced the concept of trophic (i.e., feeding) levels that are the basis of energy flow and transformation [6]. The energy flow can start with plants as primary producers as they capture energy from the sun and convert it to plant tissue. The energy captured by plants is moved to the first trophic level that is comprised of organisms that eat plants (i.e., herbivores). Organisms that eat herbivores make up the second trophic level. The third trophic level would be predators that consume the predators on the second trophic level. Although energy moves across trophic levels, the flow is not linear but cyclic. When individuals at higher trophic levels die, they are broken down by organisms called decomposers or detritivores (e.g., bacteria, fungi). The decomposers break down dead organic matter and make the nutrients available to plant species, causing the trophic cycle to begin again.

Interestingly, Lindeman found that not all the energy from one trophic level is transferred to the next. In fact, only about 10% of the energy from one trophic level flows to the next. The other 90% is used by the organisms at the lower trophic levels for their respiration and growth. Because the amount of energy is so markedly reduced as it moves up the trophic levels, there are usually three and at best four levels in an ecosystem. To obtain enough energy to grow and reproduce, top-tier species often require large ranges to roam and hunt for food. Because of this requirement, top-trophic-level species are especially vulnerable to habitat fragmentation and destruction. When ecosystems are perceived as energy flows among organisms and that available energy declines with each trophic level, resources no longer appear to be unlimited and available for use without repercussions. Instead, removing natural resources impacts the entire ecosystem by severing the energy flow that connects the species at each trophic level.

Reimagining nature as interconnected organisms living in communities challenges assumptions that nature exists for human use and our actions are isolated events. In reality, those actions reverberate through the connections among species in communities and can have broad and long-range repercussions. Understanding that the connections among organisms include our species, presages that the repercussions of our actions ultimately will affect us. Climate change is a reminder that our relationship with nature determines our quality of life and very survival on this planet along with those of generations to come.

So, what is our responsibility as individuals in addressing climate change? This is difficult to answer. Climate change has been caused by the aggregation of systems producing individual actions that were not intended to produce such an outcome. Climate change is the result of innumerable contributions of individuals over time, but no individual's contribution to greenhouse gases in the atmosphere is significant enough to cause the climate to change on a global level. An average American emits about one two-billionth (17 metric tons) of annual global emissions of CO_2 [7]. If no one is directly culpable for our changing climate, do we face the possibility that the global environment may be destroyed, yet no one will be responsible [8]?

If individuals contribute little to the overall amount of greenhouse gases in the atmosphere, how can individuals be responsible and how would changes in lifestyle to reduce greenhouse emissions have any effect? Changes we make in reducing carbon emissions will barely affect global temperatures in our lifetimes. The reductions in CO_2 levels we can make as individuals seem negligible in light of the global impact of countries with little interest or the economic means to reduce their contributions to greenhouse gases (see ref. [7]). Climate change falls into the paradox of small effects where a set of seemingly insignificant actions can collectively bring about significant harm [9].

To address climate change at an individual level requires broadening our sense of responsibility beyond our own group or borders. Globalization and the interconnectedness of commerce have pulled societies closer together, making them more interdependent and the consequences of our actions more far reaching. Still, lifestyles are hard to change. Sunk financial and emotional costs along with a sense of individual identity from how and where we live make changes in lifestyle difficult. A tension between the rights of an individual and those of the community arise especially in societies where individuality is valued.

So, what would incentivize individuals to step outside themselves and alter their lifestyles to reduce greenhouse gases for the benefit of those they do not know or see? Perhaps the strongest incentive is that climate change will eventually impact us all. Climate models project that during the second half of the 21st century, there will be up to a 50% increase in the frequency and intensity of the hottest days in Europe and North America [10,11]. The number of related illnesses and deaths increase under conditions of extreme heat in both children and the elderly with chronic respiratory disease [12]. Those with cardiovascular disease also are at risk as ambient temperatures above of 20°C are associated with a higher incidence of heart attacks (myocardial infarction) [13]. Air pollutants especially ozone and particulate matter increase in concentration at higher temperatures. These contaminants can cause premature deaths due to a reduction in lung function and increases in respiratory symptoms (including asthma) in individuals with preexisting heart and lung disease. During the European heatwave that occurred in July and August of 2003, mortality rates increased particularly among the elderly of Northern European countries, where individuals were not acclimated to the extreme heat. In the United States, approximately 400 people die from heat-related illness each year (heat exhaustion and heat stroke) [14].

Climate change has increased the frequency of natural disasters around the globe especially in coastal areas, the western United States, southern Australia, and across Africa and Asia (see World Meteorological Organization, https://wmo.int/). Climate-related disasters are couched in economic losses and $kilometers^2$ or acres that are destroyed; however, it is individuals that are displaced by these events. Displacement occurs when homes and other living spaces are destroyed by climate-related disasters such as fires and floods. People also are displaced by slow-onset climate crises where rising temperatures, water shortages, and loss of farmland make previously inhabited areas no longer livable. In the United States, over 9 million people were displaced from climate-related disasters over the last decade. Over 3 million Americans were displaced by natural disasters such as hurricanes, floods, and fires in 2023 alone. Children are particularly vulnerable to climate-related displacement. Although children under the age of 18 comprise about 30% of the world population, they make up 40% of all people forcibly displaced by climate-related disasters [15]. On the opposite end of the age spectrum, climate-related displacement accelerates population aging in affected areas because older individuals are the least likely to relocate. The shift in demographics can change the type of infrastructure needed in communities from schools and playgrounds to healthcare facilities, accessible transportation, and expanded senior housing options. Changes in demographics, socioeconomic composition and the establishment of long lasting communities will continue to occur due to climate change. Projections indicate that by 2050 up to 1.2 billion people globally could be displaced due to climate impacts in their countries [16]. The U.N. reports that as many as 3.6 billion people, nearly half the global population, currently are highly vulnerable to climate change impacts, from droughts, floods, heat stress, fire, and food insecurity.

The displacement of individuals translates into the dissolution of communities. Areas that were safe from climate-related events and that built long standing communities may become vulnerable to climate-related disasters in the future. Shifts in areas experiencing population growth also may put more people in harms way. Over the past four decades in the United States, coastal areas and the southern and western regions of the country have seen the fastest population growth. These growing areas are the most sensitive to coastal storms, drought, heat waves, fire, and air pollution [17], which could result in the displacement of residents who moved into these regions in the hope of a better life.

The loss of community deeply impacts human emotions, as there is a loss of one's sense of place and historical continuity. With relocation, individuals can experience a loss of family traditions, local knowledge, and cultural heritage that provide meaning and rootedness [18]. When entire communities are displaced, there may be little chance of restoring the deep cultural traditions even if buildings and infrastructure are restored. A prominent example was after Hurricane Katrina hit New Orleans, Louisiana, USA, in August of 2005. Nearly 80% of the population of New Orleans (approximately 385,000) were evacuated. A year later, half of the residents had not returned. By 2022, the city's population remained more than 20% lower than its pre-storm level with a substantial percentage of the original, particularly lower-income, population never returning [19–21].

So, can changes in lifestyle to reduce our carbon footprint really slow climate change? The paradox of small effects mentioned above works in many ways. Change can come from the cumulative effects of large numbers of us working together. Studying honey bee colonies has shown me what can be accomplished by working together with each individual making small contributions. I have seen bees build comb and fill it with nectar such that 50–100 lbs of honey was collected in a few days. Each bee contributed a small part collecting nectar, producing wax, and building comb, but together the bees stored enough resources for the colony to grow and prepare for the winter. Another insight gained from watching honey bees is that their behaviors benefit the colony in the short term and help sustain it into the future. The resources gathered during honey flows will not so much be used by the foragers that collected them, but instead will feed bees that are reared and emerge weeks after the foragers have died. This social structure has contributed to the success of honey bee colonies for tens of millions of years, and a similar agenda and mindset are needed to be successful in addressing the challenges of climate change.

What we can learn from studying honey bees is that individual behaviors can impact the survival of a colony, but the collective behaviors of a colony can change environments. Collective action can create the momentum needed to push past the inertia of the status quo and reverse the course that is causing climate change. Adjusting lifestyles and investing in infrastructure needs to occur within communities to create a culture where repercussions of individual choices are considered in a larger context. These are communities willing to answer questions of "what do we value?" and "who do we value?". Do we value one another equally? Do we value those not yet born who will realize the outcomes our decisions?

Collective actions that benefit a larger whole occur in numerous animal and plant species as do the means for connection and reciprocity among individuals. In over 4 billion years of evolution, the relationships among organisms produce basic patterns that are evident in the way ecosystems are organized for sustainability. The relationships between pollinators and flowering plants are an example of the value of reciprocity in sustaining long-term relationships that can have immediate and far-reaching benefits.

The immediate benefits are that flowers provide food for the bees and while foraging, bees pollinate the flowers and initiate seed set. The far-reaching benefits of plant–pollinator relationships extend across ecosystems because a cascade of events can follow seed set. The next generation of plants born from the seeds ensure their continued presence in established stands. The seeds also can be dispersed to new locations where they can establish. The plants that are produced from the seed ensure food sources, nesting sites, and cover for wildlife in the future. Plant roots contribute to soil stability and the establishment of soil organisms needed for nutrient recycling. In this way, the reciprocal relationship between the bees and plants is not just for the immediate benefits of both species, but for multiple species now and in the future.

Because multiple systems contribute to climate change, numerous collective actions are required that share a common goal of reducing carbon emissions into the environment. Collective actions usually begin with small groups, but they can grow and become larger grassroot movements that affect social norms at national and global levels. There is precedence for the power of collective action that begins with small groups of dedicated individuals. The Civil Rights Movement and women's suffrage in the U.S., the German Peace Movement to prevent the deployment of nuclear weapons in western Europe, and more recently the Global Climate Strike (Greta Thunberg) in Sweden to demand climate action all began with numerous small groups that came together to act collectively to drive systemic change. Collective behaviors and choices have resulted in the harms of climate change so there is a collective responsibility particularly to lower-income regions and communities that are experiencing the most damaging effects. There also is an intergenerational responsibility because it is very likely that future generations will face much greater climate burdens and dangers than us [22].

Many connections among individuals and institutions needed to reduce global warming already are in place. There is substantial growth in the use of solar and wind energy and other renewable sources (see https://www.climatecentral.org/climate-matters/solar-and-wind-2026). New technologies to store renewable energy and capture and remove carbon from the atmosphere have enormous economic potential on a global scale. There also are tens of thousands of organizations around the world working on climate change issues. Participants span generations (e.g., Re-Earth Initiative, Fridays for Future, Kids Against Plastic) and include non-government organizations (NGOs), civic leaders, artists, scientists, and corporations.

The Paris Climate Change Agreement adopted in December, 2015, is an international treaty that brings together countries to enact reductions in greenhouse gas emissions (mitigation) and investment in climate resilience (adaptation) [23]. The overarching goal of the Agreement is to limit the global average temperature increase to 1.5°C above pre-industrial levels. What is needed to carry out these approaches is the will and investment to move past fossil fuels and create new visions for lifestyles that resemble natural ecosystems with cyclic energy flows from renewable sources. Changes also are needed in infrastructure, transportation, agriculture, and everyday practices so that they are better suited to the conditions we live in now and can expect in the future. A sense of urgency is needed to make changes that help accomplish the goals of the Paris Climate Change Agreement. The U.N. reports that as many as 3.6 billion people currently are highly vulnerable to climate change impacts from drought, floods, heat stress, fire and food insecurity. We are long past the natural carrying capacity of the atmosphere to absorb greenhouse gases, so there are no harmless emissions [24].

Our earliest ancestors formed communities primarily to increase their chances of survival. Raising our young that take years to mature, securing food and sufficient calories to maintain activities of daily life, building shelters for protection against the elements and fending off predators were more likely to be successful if done collaboratively within a group. Small nomadic groups occured first but later became larger and more permanent communities with the development of agriculture (the Neolithic Period,c. 10,000 BCE). Permanent shelter in villages and food security supported population growth and the formation of settlements (villages). Additionally, the bonds created among the people living in small groups and later in villages provided a sense of safety and belonging that we know are fundamental human needs and motivation and are essential for maintaining our physical and mental health (see ref. [25]).

The complex interactions among individuals that impact our physical and mental health and occur at multiple levels across society can be captured in social ecological models. Like ecological models of natural ecosystems, social ecological models include the effects of environment. The models also include the complexity of relationships that impact individual health, including those with family and peers, neighborhoods and social groups, and economics, societal structures, and public policy [26].

Social ecological models have similar attributes to those of natural ecosystems including interdependence, adaptation, cycling of resources, and succession [27]. The structural bases of social ecological models and those of natural ecosystems follow similar principles. One of those principles that has been previously highlighted is that increasing diversity creates multiple connections among organisms and greater complexity. That complexity can generate stability and resilience in the face of environmental disruptions.

One way that species and genetic diversity are created in natural ecosystems is by immigration. The movement of individuals from populations outside an ecosystem also can be a rescuing mechanism for communities that are unraveling due to declining populations. Diversity through immigration can make communities more stable and resilient through connections that generate collective benefits. Mutualisms among species can develop and further ensure each other's survival [28].

An important consideration about immigration is that the benefits are predicated on the immigrating species forming connections within the community. If species migrate into communities, and those connections are not secured, species isolation can occur. The isolation not only undermines the benefits that communities might obtain from increased biodiversity, but may also cause fragmentation that threatens the stability of the community itself. Similar outcomes can occur in human societies. If new connections are made with individuals immigrating into communities, there can be an influx of culture, resources, and new ideas that create synergisms and improve the quality of life for all community members. However, immigration without connection to the social ecosystem leads to isolation as in natural ecosystems and produces alienation and perceptions of the immigrant as "the other," a permanent "outsider." With the expected increase in displacement and immigration due to climate change, it is imperative that communities know how to establish connections with newcomers at multiple levels and thereby realize the benefits of diversity to increase community stability and resilience.

The true cost of fossil fuel use and the emissions that result diffuse across time and geography. The worst impacts of climate change are felt by the least affluent communities, even though they often contributed little to creating the problem. Therefore, the spirit of the community needs to be expansive, crossing borders and political divides and beyond our lifetimes. Future generations that have no voice in climate policy decisions will be most affected by them. Perseverance also is needed. It took almost 200 years to create the state we are in, and it will take time to change it. The benefits of reducing CO_2 levels will not be realized for decades. Still, the behaviors we choose today are not to benefit us, but to the ones that follow.

Perhaps the greatest challenge to addressing climate change is to overcome our deep-seated behaviors that cause us to understand community in limited terms and see those who are not like us as 'the other'. Humans evolved living in groups that included kin and immediate social circles. We evolved prosocial behaviors that supported the functioning of our group and in return we received safety and security that could not be achieved on our own. Our strong sense of group affiliation and loyalty may not just be rooted in basic behaviors but in the evolution of our neurobiological systems and circuitries. The neuropeptide oxytocin functions as both a hormone and a neurotransmitter and has a well-documented role in reproduction and pair-bond formation (see refs in [29]). Oxytocin might also have broader functions that include social behaviors that support the functioning and safety of an individual's group. Those behaviors can include competitive reactions and noncooperation to those we perceive as threatening outsidersand rival outgroups [29].

So how do we break through our limited perceptions of community to include those beyond our immediate groups. We must find a way as this is key to solving the challenges of climate change. Perhaps a place to start is common ground we share in terms of human needs. Shared narratives and systems that promote equality and build trust can expand our collective imaginations to see humanity and its place in the natural world in more holistic terms. Global cooperation needed to address climate change requires us to rewire our behaviors and institutions, rethink the stories we tell, and expand our beliefs and incentives. We see boundaries between groups, but nature does not. Energy flows among organisms as they are all interconnected and interdependent. The natural world and our place in it can be where we might start to expand our imagination and dissolve boundaries between groups and, in the process, find a road map to lead us out of the climate crisis.

References

[1] Fouquet R. Long run trends in energy-related external costs. Ecol Econ 2011;70(12):2380–9.

[2] https://www.climate.gov/news-features/understanding-climate/climate-change-global-temperature.

[3] Parmesan C. Ecology and evolutionary responses to recent climate change. Annu Rev Ecol Evolut Syst 2006;37:637–69.

[4] Tansley AG. The use and abuse of vegetational concepts and terms. Ecology 1935;16(3):284–307.

[5] Mougi A, Kondoh M. Diversity of interaction types and ecological community stability. Science 2012;337(6092):349–51.

[6] Lindeman RL. The trophic-dynamic aspect of ecology. Ecology 1942;23(4):399–417.

[7] Fragnière A. Climate change and individual duties. Clim Change 2016;7(6):798–814.

[8] Jamieson D. Ethics, public policy, and global warming. Science, Technology, & Human Values 1992;17(2):139–53.

[9] Cullity G. 8 - Acts, omissions, emissions. In: Moss J, editor. Climate change and justice. Cambridge: Cambridge University Press; 2015. p. 148–64. https://doi.org/10.1017/CBO9781316145340.009.

[10] Duffy PB, Tebaldi C. Increasing prevalence of extreme summer temperatures in the U.S. Clim Change 2012;111:487–95.

[11] Ganguly AR, Steinhaeuser K, Erickson III DJ, Branstetter M, Parish ES, Singh N, et al. Higher trends but larger uncertainty and geographic variability in 21st century temperature and heat waves. Proc Natl Acad Sci U S A 2009;106(37):15555–9.

[12] Medina-Ramón M, Zanobetti A, Cavanagh DP, Schwartz J. Extreme temperatures and mortality: assessing effect modification by personal characteristics and specific cause of death in a multi-city case-only analysis. Environ Health Perspect 2006;114(9):1331.

[13] Bhaskaran K, Armstrong B, Hajat S, Haines A, Wilkinson P, Smeeth L. Heat and risk of myocardial infarction: hourly level case-crossover analysis of MINAP database. BMJ 2012;345:e8050.

[14] Francini M, Mannucci PM. Impact on human health of climate changes. Eur J Intern Med 2015;26:1–5.

[15] Rai AG. Climate displacement can permanently hamper children's education and imperil future prospects. Migration Policy Institute; 2025. https://www.migrationpolicy.org/article/climate-displacement-education.

[16] Hauer ME, Jacobs SA, Kulp SA. Climate migration amplifies demographic change and population aging. Proc Natl Acad Sci U S A 2024;121(3). p. e2206192119.

[17] USGCRP. In: Melillo JM, Richmond TC, Yohe GW (Eds.), Climate Change Impacts in the United States: The Third National Climate Assessment, in United States Global Change Research Program; 2014.
[18] Pearson J, Jackson G, McNamara KE. Climate-driven losses to knowledge systems and cultural heritage: a literature review exploring the impacts on Indigenous and local cultures. Anthropocene Rev 2021;10(2):343–66.
[19] North America. Climate change will drive domestic migration across the United States. New America: https://www.newamerica.org/insights/climate-migrations-impact-on-housing-security/climate-change-will-drive-domestic-migration-across-the-united-states/.
[20] Gutmann MP, Field V. Katrina in historical context: environment and migration in the US. Popul Environ 2010;31(1):3–19.
[21] Sastry N. Displaced New Orleans residents in the aftermath of Hurricane Katrina: results from a pilot survey. Organ Environ 2010;22(4):395–409.
[22] Hormio S. Collective responsibility for climate change. WIREs Clim Change 2022;14:e830.
[23] Fankhauser S. Adaptation to climate change. Annu Rev Resour Econ 2017;9:209–30.
[24] Broome J. Against denialism. Monist 2019;102(1):110–29.
[25] Michalski CA, D L, Helliwell JF, Goel V, Rosella LC. Relationship between sense of community belonging and self-rated health across life stages. SSM Popul Health 2020;12:100676.
[26] Richard L, Gauvin L, Raine K. Ecological models revisited: their uses and evolution in health promotion over two decades. Annu Rev Public Health 2011;32(1):307–26.
[27] Lounsbury DW, Mitchell SG. Introduction to special issue on social ecological approaches to community health research and action. Am J Commun Psychol 2009;44:213–20.
[28] Galla T. How migration may stabilize the diversity of ecosystems. Physics 2024;17:123.
[29] De Dreu CKW. Oxytocinergic circuitry motivates group loyalty. In: Mikulincer M, Shaver PR, editors. Mechanisms of social connection: From brain to group. American Psychological Association; 2014. p. 391–407. https://doi.org/10.1037/14250-022.

Index

Note: Page numbers followed by "*f*" refer to figures.

I

J

K

L

M

N

O

P

Printed and bound by CPI Group (UK) Ltd, Croydon, CR0 4YY
15/09/2026
02217075-0002